STUDENT GUIDE AND WORKBOOK

———— Seventh Edition ————

BASIC CHEMISTRY

STUDENT GUIDE AND WORKBOOK

Seventh Edition

BASIC CHEMISTRY

G. WILLIAM DAUB
Department of Chemistry
Harvey Mudd College

WILLIAM S. SEESE
Department of Chemistry
Casper College

PRENTICE HALL, UPPER SADDLE RIVER, NJ 07458

Production Editor: *Lisa Protzmann*
Acquisitions Editor: *Ben Roberts*
Supplement Acquisitions Editor: *Ashley Scattergood*
Production Coordinator: *Alan Fischer*

 ©1996 by Prentice-Hall, Inc.
Simon & Schuster / A Viacom Company
Upper Saddle River, New Jersey 07458

Printed in the United States of America

10 9 8 7 6 5 4 3 2 1

ISBN: 0-13-378530-0

Prentice-Hall International (UK) Limited, *London*
Prentice-Hall of Australia Pty. Limited, *Sydney*
Prentice-Hall Canada Inc., *Toronto*
Prentice-Hall Hispanoamericana, S.A., *Mexico*
Prentice-Hall of India Private Limited, *New Delhi*
Prentice-Hall of Japan, Inc., *Tokyo*
Simon & Schuster Asia Pte. Ltd., *Singapore*
Editora Prentice-Hall do Brasil, Ltda., *Rio de Janeiro*

Contents

PREFACE TO THE STUDENT GUIDE AND WORKBOOK

To the Student: The **STUDENT GUIDE AND WORKBOOK**, a supplement to **BASIC CHEMISTRY**, and BASIC CHEMISTRY: AN ALTERNATE EDITION, Seventh Edition, by G. William Daub and William S. Seese, is closely tied to **BASIC CHEMISTRY**. It is for students who might have difficulty studying chemistry and need an additional aid. Each chapter in this book contains the following:

(1) Selected Topics. These topics give a summary of a particular section in your text emphasizing the important points in that section.

(2) Problems. This book gives problems similar to those in your text. You will find detailed solutions following the problem, but do not look at the solution to the problem before you attempt to solve it. Write your solution next to the problem in the space provided in the book.

(3) Quizzes. Quizzes similar to the ones you will be given in class are at the end of each chapter. Take these quizzes. The answers and solutions are at the end of the quiz.

(4) Multiple Choice Review Exams. These review exams cover three or four chapters. These questions may be similar to ones you have on your class exams. Take these exams. The answers are given in parentheses next to the question.

In using this book, you should:

(1) STUDY the material in your text.

(2) READ the Selected Topics in this book.

(3) DO the problems in this book. If you have difficulty doing these problems, the book will refer you to your text to study various sections, examples, or study exercises.

(4) TAKE the chapter quizzes and multiple choice review exams.

G. William Daub
Department of Chemistry
Harvey Mudd College
Claremont, California 91711

William S. Seese
Emeritus
Department of Chemistry
Casper College
Casper, Wyoming 82601

CHAPTER 1

INTRODUCTION TO CHEMISTRY

This chapter introduces you to science and chemistry. It gives you an introduction to the scientific method, branches of chemistry, and a brief history of chemistry. You should note that people of all nationalities and ethnic groups have contributed to the development of chemistry.

SELECTED TOPICS

1. The scientific method. The three steps in the scientific method are:
   1. experimentation - collection of facts and data by observing natural events under controlled conditions.
   2. hypothesis - a tentative theory to explain the results of experimentation. Examining and correlating these results to propose a hypothesis.
   3. further experimentation - confirming or rejecting the hypothesis and proposing a scientific theory or law if possible.
   See Section 1-1 in your text.

2. The branches of chemistry. Key terms in the five branches of chemistry are:
   1. organic chemistry - study of substances containing carbon (symbol C).
   2. inorganic chemistry - study substances other than those containing carbon.
   3. analytical chemistry - study of what is in a sample (qualitative) and how much of it is there (quantitative).
   4. physical chemistry - study of the structure of substances, how fast (rate) they change (kinetics), and the role of heat in a chemical process (thermodynamics).
   5. biochemistry - study of chemical reactions in living organisms.
   See Section 1-2 in your text.

PROBLEM 1-1

You are considering buying a new car. You look at the style of a number of cars and the style of the Little Jewel appeals to you. You go to the car dealer and find that the Little Jewel is in your price range. Next, you take a test drive in the Little Jewel on the highway and in town and find that its performance is excellent. Which of the steps is experimentation? Hypothesis formation? Further experimentation?

ANSWERS TO PROBLEM 1-1

Experimentation: Look at the design of a number of cars.

Hypothesis: The <u>Little Jewel</u> is the car for you.

Further Experimentation: Text drive the <u>Little Jewel</u> on the highway and in town.

If your answers are correct, proceed to Problem 1-2. If your answers are incorrect, review Section 1-1 (Study Exercise 1-1) in your text.

Proceed to Problem 1-2.

PROBLEM 1-2

Listed below are research interests of international chemists. Using the definitions of the branches of chemistry, classify these research papers as organic, inorganic, analytical, physical, or biochemical.

(a) Reactions of the carbonyl group of aldehydes and ketones

[-CHO and $\begin{matrix} \backslash \\ C = O \\ / \end{matrix}$, respectively] - Reynold C. Fuson

(a) _____

(b) Thermodynamics and thermochemistry, physical chemistry of petroleum and its components, and critically evaluated numerical data from science and technology - Frederick D. Rossini

(b) _____

(c) Defects in glucose utilization in the brain of mice infected with poliomyelitis - Efrain Racker

(c) _____

(d) Means by which the genetic material of the living cell directs the synthesis of cell proteins - Marshall W. Nirenberg

(d) _____

ANSWERS TO PROBLEM 1-2

(a) Organic - symbol C (carbon)

(b) Physical

(c) Biochemical - living organisms (brain)

(d) Biochemical - living organisms (cells)

If your answers are correct, proceed to Problems 1-3. If your answers are incorrect, review Section 1-2 (Study Exercise 1-2) in your text.

Proceed to Problem 1-3.

PROBLEM 1-3

The following is a list of research papers published in various scientific journals. Using the definitions of the branches of chemistry, classify these research papers as organic, inorganic, analytical, physical, or biochemical.

(a) "Cation-Anion Combination Reactions. 23. Solvent Effects on Rates and Equilibria of Reactions" - Calvin D. Ritchie

        (a) _____

(b) "Energy Disposal in the Photofragmentation of Pyruvic Acid ($C_3H_4O_3$) in the Gas Phase" - Robert N. Rosenfeld and Brad Weiner

        (b) _____

(c) "Ruthenium [Ru] and Osmium [Os] Thiolate Compounds" - Stephen A. Koch and Michelle Millar

        (c) _____

(d) "Histone Synthesis by Lymphocytes [blood cells] in G_0 and G_1 - William I. Waithe, Jean Renaud, Paul Nadeau, and Dominick Pallotta.

        (d) _____

ANSWERS TO PROBLEM 1-3

(a) Physical - rates (how fast?)

(b) Organic - carbon (symbol C)

(c) Inorganic - no carbon (other substances)

(d) Biochemical - living systems (blood cells)

If your answers are correct, you have completed this chapter. If your answers are incorrect, re-read Section 1-2 in your text.

These problems conclude the chapter on introduction to chemistry.

Now take the sample quiz to see if you have mastered the material in this chapter.

SAMPLE QUIZ

Quiz #1

1. You are having difficulty getting your dirty white clothes, white. You decide to use a bleaching agent with your laundry. You add a little of a bleaching agent to your laundry and find that the clothes appear to look whiter, but not really white. You repeat the addition of the bleaching agent to the washing next week, but this time record the exact amount of the bleaching agent you added. Now the clothes appear really white. Which of the steps is experimentation? Hypothesis formation? Further experimentation?

2. Listed below are the research interests of some international chemists. Using the definitions of the branches of chemistry, classify these research interests as organic, inorganic, analytical, physical, or biochemical.

   (a) High temperature properties of materials; thermal electrical, expansion properties at high temperatures; structural interpretation - Paul Wagner

   (b) Bacterial cell wall structure, synthesis and function- Bruce A. McFadden

   (c) Boron hydrides (BH_3) - Thomas C. Bissat

3. The following is a list of research papers published in various scientific journals. Using the definitions of the branches of chemistry, classify these research papers as organic, inorganic, analytical, physical, or biochemical.

   (a) "Tetramethylene Dications $(CH_2)_4{}^{2+}$: Are the Norbornadiene Dication, the Pagondane Dication, and Related Systems Aromatic?" - Rainer Herges, Paul von Ragué Schleyer, Michael Schindler, and Wolf-Dieter Fessner.

   (b) "Structures of Ag_2X^+ and Cu_2X^+ Ions: Comparison of Theoretical Predictions with Experimental Results from Mass Spectrometry / Mass Spectrometry" - R. A. Flurer and K. L. Busch

   (c) "How Can the Solvent Affect Enzyme Enantioselectivity?" - Paul A. Fitzpatrick and Alexander M. Klibanov

Answers for Quiz #1

1. Experimentation: Adding a bleaching agent to your dirty
   white clothes.
   Hypothesis: The bleaching agent makes the dirty clothes
   whiter.
   Further Experimentation: Repeat the washing with various
   amounts of the bleaching agent and see if the dirty clothes
   continue to come out white.

2. (a) Physical; (b) Biochemical; (c) Inorganic

3. (a) Organic; (b) Physical; (c) Biochemical

CHAPTER 2

MEASUREMENTS

This chapter provides the basis of our problem solving technique, <u>dimensional analysis</u> (the factor-unit method). You should master this technique as we will be using it throughout the book. Also in this chapter we will review some basic mathematical skills. Then, we will apply these skills and dimensional analysis to solve problems involving the metric system, temperature, density, and specific gravity.

SELECTED TOPICS

1. <u>Matter</u> is anything that has mass and occupies space. To define this matter we can measure its mass, weight, length, volume, temperature, and density. See Section 2-1 in your text.

2. We measure this matter by using the metric system. This system is based on units that are multiples of 10, 100, 1000, etc. This system has as its basic units the gram (g, mass), the liter (or litre, L, volume), the meter (or metre, m, length). Prefixes to these basic units relate the unit to the basic unit. For example, the prefix "kilo" represents 1000 of a basic unit, such as 1000 g = 1 kg (kilogram). Table 2-1 gives the relations of the basic units of mass, volume, and length to these prefixes. You must learn these units and their equivalents in order to work problems. See Section 2-2 in your text.

3. The temperature of matter is measured by three different temperature scales. They are:

    (1) the Fahrenheit scale - $^{\circ}F$
    (2) the Celsium scale - $^{\circ}C$
    (3) the Kelvin scale - K (no degree sign)
    See Section 2-3 in your text.

4. Measuring devices used in chemistry to measure matter are: balances to measure mass; scales to measure weight; cylinders, burets, and pipettes to measure volume. See Section 2-4 in your text.

5. Significant digits are used to determine the precision (exactness) of a measurement. <u>Significant digits</u> (figures) are digits in a measurement that are known to be precise, along with a final digit about which there is some uncertainity. To determine the number of significant digits in a number, we follow the rules given in Section 2-5 in your text. Nonzero digits and confined zeros are significant. Leading zeros are not significant. Trailing zeros are significant only if the number (a) contains a decimal point or (b) contains an overbar. You must know these rules in order to identify significant digits. See Section 2-5 in

your text.

6. These significant digits affect the rounding off of answers after adding, subtracting, multiplying, and dividing numbers. Rules for rounding off are given in Section 2-6. In addition and subtraction, the answer must not contain a smaller place than the number with the smallest place. In multiplication and division, the answer must not contain any more significant digits than the least number of significant digits in the numbers used in the multiplication or division. Section 2-6 in your text gives various examples.

7. Exponential notation expresses a number as a product of two numbers, one a decimal and the other a power of 10. Scientific notation is a special form of exponential notation with the decimal point containing only one nonzero digit to the left of the decimal point. Section 2-7 in your text gives various examples of mathematical operations involving exponential notation.

8. We use dimensional analysis (the factor-unit method to solve problems throughout this text. It is based on the fact that measured quantities have units and we can use these units to solve problems in a logical manner. We use dimensional analysis to solve the following types of problems:

   (1) metric conversions
   (2) temperature conversions
   (3) density conversions

   (1) In metric conversions you must know the meaning of the prefixes given in Table 2-1 in your text

   (2) In temperature conversions you must know or be able to derive the equations for conversion from ^{O}F to ^{O}C and vice versa and the equation for conversion from ^{O}C to K. These are equations 2-2, 2-3, and 2-4 in your text.

   (3) In density conversions, you must know the meaning of density, that is, mass of a substance occupying a unit volume.

   Section 2-8 in your text gives various examples of metric, temperature, and density conversions.

9. Specific gravity is related to density. The specific gravity of a substance is the density of the substance divided by the density of some substance taken as a standard. For the specific gravity of liquids and solids, water at $4^{O}C$ is the standard with a density of 1.00 g/mL in the metric system.

$$\text{specific gravity} = \frac{\text{density of substance}}{\text{density of water at } 4^{\circ}C}$$

Specific gravity has has no units. To determine the density of a substance in the metric system given the specific gravity, we multiply by 1.00 g/mL (the density of water at $4^{\circ}C$).

density of substance
   $= \text{specific gravity} \times \text{density of water at } 4^{\circ}C$

The specific gravity is often expressed as follows:

$$\text{sp gr} = 1.26^{20^{\circ}/4} \text{ of glycerine}$$

The 20° refers to the temperature in $^{\circ}C$ at which the density of the glycerine was measured and the 4 refers to the temperature in $^{\circ}C$ at which the density of water was measured. Various examples involving specific gravity are given in Section 2-9 in your text.

PROBLEM 2-1

Calculate the density of a piece of metal that has a mass of 28 g and occupies a volume of 3.5 mL.

ANSWER TO PROBLEM 2-1

8.0 g/mL

$$\frac{28 \text{ g}}{3.5 \text{ mL}} = 8.0 \text{ g/mL}$$

If your answer is correct, proceed to Problem 2-2. If your answer is incorrect, review Section 2-2 (Example 2-1 and Study Exercise 2-1), in your text.

Proceed to Problem 2-2.

PROBLEM 2-2

Determine the number of significant digits in the following numbers:

(a) 0.08   (a) _____

(b) 0.095   (b) _____

(c) 1.3   (c) _____

(d) 1.005   (d) _____

(e) 0.0630   (e) _____

(f) 33,3̄00   (f) _____

8

ANSWERS TO PROBLEM 2-2

(a)   1

(b)   2

(c)   2

(d)   4

(e)   3

(f)   4

If your answers are correct, proceed to Problem 2-4. If your answers are incorrect, review Section 2-5 (Example 2-2 and Study Exercise 2-2) in your text.

Proceed to Problem 2-3.

PROBLEM 2-3

Determine the number of significant digits in the following numbers:

(a)   305   (a) _____

(b)   3.05   (b) _____

(c)   0.025   (c) _____

(d)   20.05   (d) _____

(e)   8.0500   (e) _____

(f)   13,$\overline{0}$00   (f) _____

ANSWERS TO PROBLEM 2-3

(a)   3

(b)   3

(c)   2

(d)   4

(e)   5

(f)   3

If your answers are correct, proceed to Problem 2-4. If your answers are incorrect, re-read Section 2-5 in your text. Re-work Problems 2-2 and 2-3.

Proceed to Problem 2-4.

PROBLEM 2-4

Round off the following numbers to <u>three</u> significant digits.

(a)	1.8392	(a)	_____
(b)	0.053952	(b)	_____
(c)	12.59	(c)	_____
(d)	2.3914	(d)	_____
(e)	0.92640	(e)	_____
(f)	87.250	(f)	_____

ANSWERS TO PROBLEM 2-4

(a) 1.84

(b) 0.0540

(c) 12.6

(d) 2.39

(e) 0.926

(f) 87.2

If your answers are correct, proceed to Problem 2-6. If your answers are incorrect, review Section 2-6, Rounding Off (Example 2-3 and Study Exercise 2-3), in your text.

Proceed to Problem 2-5.

PROBLEM 2-5

Round off the following numbers to <u>four</u> significant digits.

(a)	1.8953	(a)	_____
(b)	0.30756	(b)	_____
(c)	5.3399	(c)	_____
(d)	2.23650	(d)	_____
(e)	14.752	(e)	_____
(f)	43.2350	(f)	_____

ANSWERS TO PROBLEM 2-5

(a)   1.895

(b)   0.3076

(c)   5.340

(d)   2.236

(e)   14.75

(f)   43.24

If your answers are correct, proceed to Problem 2-6. If your answers are incorrect, re-read Section 2-6, Rounding Off, in your text. Re-work Problems 2-4 and 2-5.

Proceed to Problem 2-6.

PROBLEM 2-6

Perform the indicated operation and express your answer to the proper number of significant digits.

(a)   35.27 + 3.198 + 102.2   (a) _____

(b)   239.90 - 25.983   (b) _____

(c)   0.85 x 327   (c) _____

(d)   3$\overline{00}$/25.5   (d) _____

ANSWERS TO PROBLEM 2-6

(a)   140.7

(b)   213.92

(c)   280

(d)   11.8

If your answers are correct, proceed to Problem 2-8. If your answers are incorrect, review Section 2-6, Simple Math (Example 2-4 and Study Exercise 2-4), in your text.

Proceed to Problem 2-7.

PROBLEM 2-7

Perform the indicated operation and express your answer to the proper number of significant digits.

(a)   3.58 + 10.465 + 0.6   (a) _____

(b) 12.835 - 3.60   (b) _____

(c) 0.03 x 0.589   (c) _____

(d) 9.57/5.183   (d) _____

ANSWERS TO PROBLEM 2-7

(a) 14.6

(b) 9.24

(c) 0.02

(d) 1.85

If your answers are correct, proceed to Problem 2-8. If your answers are incorrect, re-read Section 2-6, Simple Math, in your text. Re-work Problems 2-6 and 2-7.

Proceed to Problem 2-8.

PROBLEM 2-8

Express the following numbers in scientific notation to three significant digits.

(a) 789.2   (a) _____

(b) 3002467   (b) _____

(c) 0.00265   (c) _____

(d) 1,250   (d) _____

(e) 0.00100   (e) _____

(f) 0.200   (f) _____

ANSWERS TO PROBLEM 2-8

(a) 7.89×10^2

(b) 3.00×10^6

(c) 2.65×10^{-3}

(d) 1.25×10^3

(e) 1.00×10^{-3}

(f) 2.00×10^{-1}

If your answers are correct, proceed to Problem 2-9. If your answers are incorrect, review Section 2-7, Scientific Notation, (Example 2-5, 2-6, and 2-7 and Study Exercises 2-5 and 2-6) in your text.

Proceed to Problem 2-9.

PROBLEM 2-9

Carry out the operations indicated on the following exponential numbers:

(a) $(1.2 \times 10^2) + (4.0 \times 10^2)$ (a) _____

(b) $(3.00 \times 10^3) + (2.0 \times 10^2)$ (b) _____

(c) $(6.70 \times 10^2) - (2.4 \times 10^1)$ (c) _____

(d) $(2.49 \times 10^5) \times (1.10 \times 10^{-3})$ (d) _____

(e) $(3.41 \times 10^{-5}) \times (3.41 \times 10^{-2})$ (e) _____

(f) $(4.90 \times 10^6) / (2.45 \times 10^8)$ (f) _____

(g) $(7.53 \times 10^{-3}) / (8.52 \times 10^{-5})$ (g) _____

(h) $\sqrt{9.00 \times 10^8}$ (h) _____

(i) $\sqrt{4.70 \times 10^{-5}}$ (i) _____

(j)   raise 2.25×10^3 to the second power   (j) _____

(k)   raise 1.42×10^4 to the third power   (k) _____

ANSWERS TO PROBLEM 2-9

(a) 5.2×10^2

(b) 3.20×10^3

(c) 6.46×10^2

(d) 2.74×10^2

(e) 1.16×10^{-6}

(f) 2.00×10^{-2}

(g) 8.84×10^1

(h) 3.00×10^4

(i) 6.86×10^{-3} Change 4.70×10^{-5} to an even exponent--

                        47.0×10^{-6}. Then take the square root

                        of 47.0×10^{-6}.

(j) 5.06×10^{6}

(k) 2.86×10^{12}

If your answers are correct, proceed to Problem 2-10. If your answers are incorrect, review Section 2-7, Mathematical Operations Involving Exponential Notation (Examples 2-8, 2-9, 2-10, and 2-11 and Study Exercises 2-7 and 2-8) in your text.

Proceed to Problem 2-10.

PROBLEM 2-10

Carry out each of the following conversions showing a solution set-up. (Express your answer in scientific notation.)

(a) 1.25 mL to L   (a) _____

(b) 32.5 kg to mg   (b) _____

(c) 0.95 cm to m   (c) _____

(d) 75 μL to L   (d) _____

(e) 97 pg to ng   (e) _____

ANSWERS AND SOLUTIONS TO PROBLEM 2-10

(a) 1.25×10^{-3} L

$$1.25 \text{ mL} \times \frac{1 \text{ L}}{1000 \text{ mL}} = 1.25 \times 10^{-3} \text{ L}$$

(b) 3.25×10^{7} mg

$$32.5 \text{ kg} \times \frac{1000 \text{ g}}{1 \text{ kg}} \times \frac{1000 \text{ mg}}{1 \text{ g}} = 3.25 \times 10^{7} \text{ mg}$$

(c) 9.5×10^{-3} m

$$0.95 \text{ cm} \times \frac{1 \text{ m}}{100 \text{ cm}} = 9.5 \times 10^{-3} \text{ m}$$

14

(d) 7.5×10^{-5} L

$$75 \; \cancel{\mu L} \times \frac{1 \text{ L}}{1,000,000 \; \cancel{\mu L}} = 7.5 \times 10^{-5} \text{ L} \quad \text{or}$$

$$75 \; \cancel{\mu L} \times \frac{1 \text{ L}}{10^{6} \; \cancel{\mu L}} = 75 \times 10^{-6} \text{ L or } 7.5 \times 10^{-5} \text{ L}$$

(e) 9.7×10^{-2} ng

$$97 \; \cancel{pg} \times \frac{1 \; \cancel{g}}{10^{12} \; \cancel{pg}} \times \frac{10^{9} \text{ ng}}{1 \; \cancel{g}} = 97 \times 10^{-3} \text{ ng} = 9.7 \times 10^{-2} \text{ ng}$$

$$\text{or} \quad 97 \; \cancel{pg} \times \frac{1 \text{ ng}}{10^{3} \; \cancel{pg}} = 97 \times 10^{-3} \text{ ng} = 9.7 \times 10^{-2} \text{ ng}$$

If your answers are correct, proceed to Problem 2-12. If your answers are incorrect, review Section 2-8, Metric Conversions (Examples 2-12, 2-13, 2-14, 2-15, and 2-16 and Study Exercise 2-9), in your text.

Proceed to Problem 2-11.

PROBLEM 2-11

Carry out each of the following conversions showing a solution set-up. (Express your answer in scientific notation.)

(a) 1.37 L to mL   (a) _____

(b) 3500 mg to kg   (b) _____

(c) 5.7 ng to pg   (c) _____

ANSWERS AND SOLUTIONS TO PROBLEM 2-11

(a) 1.37×10^{3} mL

$$1.37 \; \cancel{L} \times \frac{1000 \text{ mL}}{1 \; \cancel{L}} = 1.37 \times 10^{3} \text{ mL}$$

15

(b) 3.5×10^{-3} kg

$$3500 \text{ mg} \times \frac{1 \text{ g}}{1000 \text{ mg}} \times \frac{1 \text{ kg}}{1000 \text{ g}} = 3.5 \times 10^{-3} \text{ kg}$$

(c) 5.7×10^{3} pg

$$5.7 \text{ ng} \times \frac{1 \text{ g}}{10^{9} \text{ ng}} \times \frac{10^{12} \text{ pg}}{1 \text{ g}} = 5.7 \times 10^{3} \text{ pg}$$

or $5.7 \text{ ng} \times \frac{10^{3} \text{ pg}}{1 \text{ ng}} = 5.7 \times 10^{3} \text{ pg}$

If your answers are correct, proceed to Problem 2-12. If your answers are incorrect, review Section 2-8, Metric Conversions, in your text. Re-work Problems 2-10 and 2-11.

Proceed to Problem 2-12.

PROBLEM 2-12

Add the following lengths:

1.050000 km, 250.1 cm, 3.850 m, 45$\overline{00}$ mm.

What is the total length in meters?

ANSWER TO PROBLEM 2-12

1060.851 m

$$1.050000 \text{ km} \times \frac{1000 \text{ m}}{1 \text{ km}} = 1050.000 \text{ m}$$

$$250.1 \text{ cm} \times \frac{1 \text{ m}}{100 \text{ cm}} = 2.501 \text{ m}$$

$$3.850 \text{ m} = 3.850 \text{ m}$$

$$45\overline{00} \text{ mm} \times \frac{1 \text{ m}}{1000 \text{ mm}} = \frac{4.500 \text{ m}}{1060.851 \text{ m}}$$

If your answer is correct, proceed to Problem 2-14. If your answer is incorrect, review Sections 2-8, Metric Conversions, in your text.

16

Proceed to Problem 2-13.

PROBLEM 2-13

Add the following masses: 1.500 g, 3025 mg, 0.850000 kg.
What is the total mass in grams?

_____

ANSWER TO PROBLEM 2-13

854.525 g

$$1.500 \text{ g} \qquad\qquad = \quad 1.500 \text{ g}$$

$$3025 \text{ mg} \times \frac{1 \text{ g}}{1000 \text{ mg}} \qquad = \quad 3.025 \text{ g}$$

$$0.850000 \text{ kg} \times \frac{1000 \text{ g}}{1 \text{ kg}} \qquad = \frac{850.000 \text{ g}}{854.525 \text{ g}}$$

If your answer is correct, proceed to Problem 2-14. If your
answer is incorrect, review Section 2-8, Metric Conversions,
in your text. Re-work Problems 2-12 and 2-13.

Proceed to Problem 2-14.

PROBLEM 2-14

Convert each of the following temperatures as indicated:

(a) 50°C to $^{\circ}\text{F}$ (a) _____

(b) -76°F to $^{\circ}\text{C}$ (b) _____

ANSWERS AND SOLUTIONS TO PROBLEM 2-14

(a) 122°F

$$°F = 1.8°C + 32$$
$$= 1.8(5\overline{0}) + 32$$
$$= 122°F$$

(b) -6$\overline{0}$°C

$$°C = \frac{(°F - 32)}{1.8}$$
$$= \frac{(-76 - 32)}{1.8}$$
$$= -6\overline{0}°C$$

If your answers are correct, proceed to Problem 2-15. If your answers are incorrect, review Section 2-8, Temperature Conversions (Examples 2-17, 2-18, and 2-19 and Study Exercises 2-10 and 2-11) in your text.

Proceed to Problem 2-15.

PROBLEM 2-15

Convert each of the following temperatures as indicated:

(a) -21°C to K   (a) _____

(b) 53°F to K   (b) _____

ANSWERS AND SOLUTIONS TO PROBLEM 2-15

(a) 252 K
$$K = °C + 273$$
$$= -21 + 273$$
$$= 252 \text{ K}$$

(b) 285 K
$$°C = \frac{(°F - 32)}{1.8}$$
$$= \frac{(53 - 32)}{1.8}$$
$$= 12°C \text{ (nearest degree)}$$
Then
$$K = °C + 273$$
$$= 12 + 273$$
$$= 285 \text{ K}$$

If your answers are correct, proceed to Problem 2-16. If your answers are incorrect, review Section 2-8, Temperature Conversions, in your text.

Proceed to Problem 2-16.

PROBLEM 2-16

Which is colder, $-80^{O}C$ or $-125^{O}F$?

_____

ANSWER AND SOLUTION TO PROBLEM 2-16

$-125^{O}F$ In order to answer this question, you must compare the two temperatures using the same scale. You may choose to convert $-80^{O}C$ to ^{O}F or $-125^{O}F$ to ^{O}C. Either conversion will give you the correct answer.

If you chose to convert $-80^{O}C$ to ^{O}F, the equation is:

$$
\begin{aligned}
^{O}F &= 1.8^{O}C + 32 \\
&= 1.8(-80) + 32 \\
&= -112^{O}F
\end{aligned}
$$

If you converted $-125^{O}F$ to ^{O}C, the equation is:

$$
\begin{aligned}
^{O}C &= \frac{(^{O}F - 32)}{1.8} \\
&= \frac{(-125 - 32)}{1.8} \\
&= -87^{O}C
\end{aligned}
$$

Either conversion provides the correct answer: $-125^{O}F$ is colder than $-80^{O}C$.

If your answer is correct, proceed to Problem 2-16. If your answer is incorrect, re-read Section 2-8, Temperature Conversions, in your text. Re-work Problems 2-14, 2-15, and 2-16.

Proceed to Problem 2-17.

PROBLEM 2-17

Calculate the volume in milliliters at 20°C occupied by a certain object having a density of 8.00 g/mL at 20°C and having a mass of 2.00 kg.

ANSWER AND SOLUTION TO PROBLEM 2-17

$25\overline{0}$ mL   This problem asks us to convert mass units to volume units (kg-->g-->mL). We can use the given density of the object as a conversion factor. Since density gives the relationship between mass and volume, we can say that 1 mL has a mass equal to 8.00 g, or that 8.00 g occupies a volume equal to 1 mL. We then solve for the volume by arranging the terms so that all units except those which express volume cancel out. Thus:

$$2.00 \text{ kg} \times \frac{1000 \text{ g}}{1 \text{ kg}} \times \frac{1 \text{ mL}}{8.00 \text{ g}} = 25\overline{0} \text{ mL}$$

Notice that if we set up the problem in the following manner, the units would not cancel and the answer would be meaningless.

$$2.00 \text{ kg} \times \frac{1000 \text{ g}}{1 \text{ kg}} \times \frac{8.00 \text{ g}}{1 \text{ mL}} = \frac{g^2}{mL}$$

If your answer is correct, proceed to Problem 2-18. If your answer is incorrect, review Section 2-8, Density Conversions (Examples 2-20, 2-21, 2-22, 2-23, and 2-24 and Study Exercises 2-12 and 2-13) in your text.

Proceed to Problem 2-18.

PROBLEM 2-18

Calculate the mass in kilograms of a steel ball having a density of 7.2 g/mL at 20°C and a volume of 5.0 L.

ANSWER AND SOLUTION TO PROBLEM 2-18

36 kg   We can use the given density of 7.2 g/mL as a conversion factor as we convert L --> mL --> kg. The solution is:

$$5.0 \; \cancel{L} \times \frac{1000 \; \cancel{mL}}{1 \; \cancel{L}} \times \frac{7.2 \; \cancel{g}}{1 \; \cancel{mL}} \times \frac{1 \; kg}{1000 \; \cancel{g}} = 36 \; kg$$

If your answer is correct, proceed to Problem 2-19. If your answer is incorrect, re-read Section 2-8, Density Conversions, in your text. Re-work Problems 2-17 and 2-18.

Proceed to Problem 2-19.

PROBLEM 2-19

Calculate the number of grams in 0.235 liter of acetic acid.

The specific gravity of acetic acid is $1.05^{20^\circ/4}$.

21

ANSWER AND SOLUTION TO PROBLEM 2-19

247 g   The specific gravity is $1.05^{20^{O}}/4$. Hence, in the metric system, the density is 1.00 g/mL x 1.05 = 1.05 g/mL.

$$0.235 \text{ L} \times \frac{1000 \text{ mL}}{1 \text{ L}} \times \frac{1.05 \text{ g}}{1 \text{ mL}} = 247 \text{ g}$$

If your answer is correct, proceed to Problem 2-20. If your answer is incorrect, review Section 2-9, (Examples 2-25, 2-26, and 2-27 and Study Exercises 2-14 and 2-15) in your text.

Proceed to Problem 2-20.

PROBLEM 2-20

Calculate the number of liters in 685 g of a certain organic liquid. The specific gravity of the liquid is $1.40^{20^{O}}/4$.

ANSWER AND SOLUTION TO PROBLEM 2-20

0.489 L   The specific gravity is 1.40. Hence, in the metric system, the density is 1.00 g/mL x 1.40 = 1.40 g/mL.

$$685 \text{ g} \times \frac{1 \text{ mL}}{1.40 \text{ g}} \times \frac{1 \text{ L}}{1000 \text{ mL}} = 0.489 \text{ L}$$

If your answer is correct, you have completed this chapter. If your answer is incorrect, re-read Section 2-9 in your text. Re-work Problems 2-19 and 2-20.

These problems conclude the chapter on measurements. Have you noticed that consistently using dimensional analysis helps you arrive at the correct solution to a problem?

Now take some sample quizzes to see if you have mastered the material in this chapter.

SAMPLE QUIZZES

The quizzes divide this chapter into two parts:

Quiz #2A (Sections 2-1 to 2-7) and Quiz #2B (Sections 2-8 and 2-9).

Quiz #2A (Sections 2-1 to 2-7)

1. Determine the number of significant digits in the following numbers:

    (a) 505 _____

    (b) 0.006040 _____

2. Round off the following numbers to three significant digits.

    (a) 4.7650 _____

    (b) 0.076456 _____

3. Express the following numbers in scientific notation to three significant digits.

    (a) 7,250 _____

    (b) 0.0785 _____

4. Perform the following operations and express your answer to three significant digits in scientific notation.

    (a) $6.45 \times 10^3 + 3.24 \times 10^2 =$

    (b) $3.24 \times 10^7 \times 2.45 \times 10^{-2} =$

    (c) $\dfrac{6.54 \times 10^2}{3.45 \times 10^{-5}} =$

    (d) $\sqrt{40.0 \times 10^3} =$

Answers for Quiz #2A

1. (a) 3 (b) 4

2. (a) 4.76 (b) 0.0765

3. (a) 7.25×10^3 (b) 7.85×10^{-2}

4. (a) 6.77×10^3 (b) 7.94×10^5 (c) 1.90×10^7 (d) 2.00×10^2

Quiz #2B (Sections 2-8 and 2-9)

1. Convert 3.75 kL to mL. Express your answer in scientific notation.

2. Convert 77°F to °C and K.

3. Calculate the volume in mL of $36\overline{0}$ g of a liquid whose density is 1.20 g/mL at 20°C.

4. Calculate the mass in grams of benzene having a density of 0.880 g/mL at 20°C and a volume of 1.10 L.

5. The specific gravity of ethyl ether is $0.708^{25°/4}$. Calculate the number of grams in 0.650 L of ethyl ether.

Solutions and Answers for Quiz #2B

1.   3.75 k̶L̶ x $\frac{1000 \text{ L̶}}{1 \text{ k̶L̶}}$ x $\frac{1000 \text{ mL}}{1 \text{ L̶}}$ = 3.75 x 10^6 mL

2.   77°F = $\frac{(77 - 32)}{1.8}$ °C = 25°C

     (25 + 273) K = 298 K

3. $36\overline{0}$ g̶ x $\frac{1 \text{ mL}}{1.20 \text{ g̶}}$ = $3\overline{00}$ mL

4.   1.10 L̶ x $\frac{1000 \text{ m̶L̶}}{1 \text{ L̶}}$ x $\frac{0.880 \text{ g}}{1 \text{ m̶L̶}}$ = 968 g

24

5. $1.00 \frac{g}{mL} \times 0.708 = 0.708 \frac{g}{mL}$

$0.650 \; \cancel{L} \times \frac{1000 \; \cancel{mL}}{1 \; \cancel{L}} \times \frac{0.708 \; g}{1 \; \cancel{mL}} = 46\overline{0} \; g$

CHAPTER 3

MATTER AND ENERGY

This chapter introduces you to matter and energy. We will consider the three physical states of matter; properties of matter and the changes matter undergoes; the classification of matter into compounds, elements, or mixtures; the meaning of a formula of a compound; energy and specific heat; and the division of elements into metals and nonmetals and their respective properties.

SELECTED TOPICS

1. The three physical states of matter are solids, liquids and gases. See Section 3-1 in your text.

2. Matter is divided into homogeneous and heterogeneous matter. Homogeneous matter is further divided into pure substances, homogeneous mixtures, and solutions. Pure substances are further divided into compounds and elements, both composed of atoms. Table 3-1 of your text gives the symbols of the more common elements. You must memorize the names and symbols of these 47 elements. See Section 3-2 in your text.

3. Compounds may be composed of molecules which are represented by a molecular formula. In one molecule of a compound, the subscripts represent the number of atoms of the element present. In one molecule of water (H_2O), there are two atoms of hydrogen and 1 atom of oxygen, a total of 3 atoms in one molecule of water. The law of definite proportions (constant composition) states that a given pure compound always contains the same elements in exactly the same proportions by mass. For example, exactly 1.0080 parts by mass of hydrogen combines with 7.9997 parts by mass of oxygen to form water. See Section 3-3 in your text.

4. Physical properties are properties of a substance that can be observed without a change in the composition of the sub-stance; a change in composition occurs when chemical prop-erties are observed. Color, solubility, density, and specific gravity are examples of physical properties; the rusting of iron, the electrolysis of water, and the burning of gasoline are examples of chemical properties. A change differs from a property in that a change is a conversion from one form of a substance to another while a property distinguishes one substance from another substance. The melting of ice is a physical change in that the solid (ice) changes to a liquid (water), but the substance is still the same and the liquid water can be frozen again and changed back to ice. The burning of bread in a toaster is

26

an example of a chemical change in that a black solid (carbon) forms on the surface of the bread, on burning. In chemical changes you should look for the production of a gas, heat energy given off, a color change, or the appearance of an insoluble substance. See Section 3-4 in your text.

5. All changes and transformations in nature are accompanied by changes in energy. Energy is the capacity to do work or transfer heat. Energy may be potential (possessed by a substance by virtue of its position in space or chemical composition) or kinetic (possessed by a substance by virtue of its motion). Heat energy is measured in calories or joules. Specific heat is a physical property of a substance and relates the amount amount of heat energy added to a substance, the mass of the substance, and the resulting change in temperature. See Section 3-5 in your text.

6. There are two conservation laws: conservation of energy (energy is neither created nor destroyed, but may be transformed from one form to another) and conservation of mass (mass is neither created nor destroyed and the total mass of the substances involved in a physical or chemical change remain constant). These two laws are related in Einstein's equation, $E = mc^2$ where E is the energy, m is the mass and c is the speed of light (3.00×10^8 m/s). See Section 3-6 in your text.

7. Elements are divided into metals and nonmetals based on the physical and chemical properties of the elements. The metals are to the left of the colored stair step line in the periodic table and the nonmetals are to the right. The last column of nonmetals is called the noble gases. Elements that lie on the colored stair step line are called metalloids (semimetals). (Aluminum is not included as a metalloid, but as a metal because it has mostly metallic properties.) See Section 3-7 in your text.

PROBLEM 3-1

Classify each of the following as a compound, element, or mixture:

(a) iron (Fe)   (a) _____

(b) sugar ($C_{12}H_{22}O_{11}$)   (b) _____

(c) salt (NaCl)   (c) _____

(d) iron, sugar, and salt in your body (d) _____

27

ANSWERS TO PROBLEM 3-1

(a)   element   The symbol is Fe.

(b)   compound   The formula is $C_{12}H_{22}O_{11}$.

(c)   compound   The formula is NaCl.

(d)   mixture   You are a mixture of many substances.

If your answers are correct, proceed to Problem 3-2.   If you answers are incorrect, review Section 3-2 in your text.

Proceed to Problem 3-2.

PROBLEM 3-2

Determine the number of atoms of each element; write the name of the element and the total number of atoms in the following molecules:

(a) SO_3 (sulfur trioxide)   (a) _____

_____

(b) N_2O_4 (dinitrogen tetroxide)   (b) _____

_____

(c) $HC_2H_3O_2$ (acetic acid)   (c) _____

_____

(d) $C_6H_{12}O_6$ (glucose, dextrose)   (d) _____

_____

ANSWERS TO PROBLEM 3-2

(a)   1 atom sulfur, 3 atom oxygen; 4 atoms total

(b)   2 atoms nitrogen, 4 atoms oxygen; 6 atoms total

(c)   2 atoms carbon, 4 atoms hydrogen (1 for first H and 3 in second H), 2 atoms oxygen, 8 atoms total

(d)   6 atoms carbon, 12 atoms hydrogen, 6 atoms oxygen; 24 atoms total.

If your answer are correct, proceed to Problem 3-3.   If your answers are incorrect, review Section 3-3, (Example 3-1 and Study Exercise 3-1) in your text.

Proceed to Problem 3-3.

PROBLEM 3-3

From the number of atoms of each element in one molecule of the compound, write the molecular formula of the following compounds:

(a)   dinitrogen pentoxide: 2 atoms nitrogen, 5 atoms oxygen

(a) _____

(b)   sulfur tetrachloride: 1 atom sulfur, 4 atoms chlorine

(b) _____

(c)   table sugar, sucrose: 12 atoms carbon, 22 atoms hydrogen, 11 atoms oxygen

(c) _____

(d)   urea: 1 atom carbon, 4 atoms hydrogen, 2 atoms nitrogen, 1 atom oxygen

(d) _____

ANSWERS TO PROBLEM 3-3

(a) N_2O_5 You must know the symbols before you can write the formulas.

(b) SCl_4

(c) $C_{12}H_{22}O_{11}$

(d) CH_4N_2O

If your answers are correct, proceed to Problem 3-4. If your answers are incorrect, review Section 3-3, (Example 3-2 and Study Exercise 3-2) in your text.

Proceed to Problem 3-4.

PROBLEM 3-4

The following are properties of the element palladium; classify them as physical or chemical properties:

(a)   forms sulfide when heated with sulfur   (a) _____

(b)   specific gravity $12.02^{20°/4}$ (b) _____

(c)   m. p. $1555°C$ (c) _____

(d)   forms phosphide when heated with phosphorus (d) _____

ANSWERS TO PROBLEM 3-4

(a) chemical Forms a sulfide, a new compound

(b) physical

(c) physical

(d) chemical Forms a phosphide, a new compound.

If your answers are correct, proceed to Problem 3-5. If your
answers are incorrect, review Section 3-4 (Study Exercise 3-3)
in your text.

Proceed to Problem 3-5.

PROBLEM 3-5

Classify the following changes as physical or chemical:

(a) burning trash   (a) _____

(b) compacting trash   (b) _____

(c) smashing your car's right fender against a
    telephone pole   (c) _____

(d) rusting of your car's right fender producing
    a big hole in the fender   (d) _____

ANSWERS TO PROBLEM 3-5

(a) chemical Gases, heat, and light energy are given off.
              Some of these gases can be used as a fuel to
              heat boilers and produce electricity. The ener-
              gy evolved can also be used.

(b) physical

(c) physical Maybe, you can get the fender repaired!

(d) chemical A red oxide (rust) forms, flakes off, and leaves
              a hole. The properties of the rust are com-
              pletely different from those of the metal.

If your answers are correct, proceed to Problem 3-6. If your
answers are incorrect, review Section 3-4 (Study Exercise 3-4)
in your text.

Proceed to Problem 3-6.

PROBLEM 3-6

(a)   Calculate the number of calories needed to raise the temperature of 1000 g of water from 25.0°C to 100.0°C. (specific heat of water = 1.00 cal/g °C)

(a)_____

(b) Calculate the number of kilocalories needed in the above problem (a).

(b)_____

ANSWERS AND SOLUTIONS TO PROBLEM 3-6

(a)   75,000 cal   Using the specific heat of water (1.00 cal/g . °C), the solution is as follows:

$$1.00 \ \frac{cal}{g \cdot °C} \ \times \ (100.0 - 25.0)°C \ \times \ 1000 \ g = 75,000 \ cal$$

(b)   75.0 kcal $$75,000 \ cal \times \frac{1 \ kcal}{1000 \ cal} = 75.0 \ kcal$$

If your answers are correct, proceed to Problem 3-7. If your answers are incorrect, review Section 3-5, (Examples 3-3, 3-4, 3-5, 3-6, and 3-7 and Study Exercises 3-5 and 3-6) in your text.

Proceed to Problem 3-7.

PROBLEM 3-7

Calculate the number of calories needed to raise the temperature of 0.500 kg of lead from 50.0°C to 90.0°C. The specific heat of lead is 0.0310 cal/g . °C.

ANSWER AND SOLUTION TO PROBLEM 3-7

$62\overline{0}$ cal

$$0.0310 \; \frac{cal}{g \cdot {}^oC} \; \times \; (90.0 - 50.0) \, {}^oC \times 0.500 \; kg \times \frac{1000 \; g}{1 \; kg} = 62\overline{0} \; cal$$

If your answer is correct, proceed to Problem 3-8. If your answer is incorrect, review Section 3-5 in your text.

Proceed to Problem 3-8.

PROBLEM 3-8

Calculate the amount of sodium chloride in grams if 44.0 calories of heat are absorbed when the sodium chloride is heated from $25.0\,{}^oC$ to $75.0\,{}^oC$. The specific heat of sodium chloride is 0.204 cal/g ${}^oC$.

ANSWER AND SOLUTION TO PROBLEM 3-8

4.31 g   The specific heat of sodium chloride is $\dfrac{0.204 \; cal}{1.00 \; g \cdot {}^oC}$.

Because we are interested in finding the number of grams of sodium chloride in this problem, we must use the units grams in the numerator. We do this by inverting the factor to read: $\dfrac{1.00 \; g \cdot {}^oC}{0.204 \; cal}$.

$$\frac{1.00 \; g \cdot {}^oC}{0.204 \; cal} \; \times \; \frac{44.0 \; cal}{(75.0 - 25.0)\,{}^oC} = 4.31 \; g$$

If your answer is correct, proceed to Problem 3-9. If your answer is incorrect, review Section 3-5 in your text.

Proceed to Problem 3-9.

PROBLEM 3-9

Calculate the number of joules required to raise the temperature of 325 g of silver from $42\,{}^oC$ to $72\,{}^oC$. The specific heat of silver is $\dfrac{2.34 \times 10^2 \; J}{kg \cdot K}$.

ANSWER AND SOLUTION TO PROBLEM 3-9

2.28×10^3 J   Again, it is best to start with the specific heat

of silver which in SI units is $\dfrac{2.34 \times 10^2 \text{ J}}{\text{kg} \cdot \text{K}}$.

First, convert both temperature to kelvins.

K = ^{O}C + 273	K = ^{O}C + 273
= 42 + 273	= 72 + 273
= 325 K	= 345 K

Notice that the unit we are solving for, joules, is in the numerator.   The next step is to arrange the terms in the problem in such a way that all other units cancel.

$\dfrac{2.34 \times 10^2 \text{ J}}{\text{kg} \cdot \text{K}}$ x (345 - 315)K x $\dfrac{1 \text{ kg}}{1000 \text{ g}}$ x 325 g = 2.28×10^3 J

Problem 3-10 provides additional practice in solving a specific heat problem.

PROBLEM 3-10

Calculate the specific heat of a metal in (a) cal/g ^{O}C and in (b) J/kg K if 6.00 kcal of heat energy will raise the temperature of 900 g of the metal from 45.0^{O}C to 60.0^{O}C. (Hint: 1 cal = 4.184 J).

(a) _____

(b) _____

33

ANSWERS AND SOLUTION TO PROBLEM 3-10

(a) 0.444 cal/g $\cdot ^{\circ}$C

$$\frac{6.00 \text{ kcal}}{900 \text{ g} \quad (60.0 - 45.0)^{\circ}\text{C}} \times \frac{1000 \text{ cal}}{1 \text{ kcal}} = 0.444 \text{ cal/g} \cdot ^{\circ}\text{C}$$

(b) 1.86 x 10^3 J/kg $\cdot$ K

Converting 45.0°C and 60.0°C to K, we have 318.0 K and 333.0 K, respectively.

$$\frac{6.00 \text{ kcal}}{900 \text{ g} \quad (333.0 - 318.0) \text{ K}} \times \frac{1000 \text{ g}}{1 \text{ kg}} \times \frac{1000 \text{ cal}}{1 \text{ kcal}} \times \frac{4.184 \text{ J}}{1 \text{ cal}}$$

$$= 1.86 \times 10^3 \text{ J/kg} \cdot \text{K}$$

If your answers are correct, your have completed this chapter. If your answers are incorrect, re-read Section 3-5 and re-work Example 3-3, 3-4, 3-5, 3-6, and 3-7 and Study Exercises 3-5 and 3-6 in your text.

These problems conclude the chapter on matter and energy.

Now take the sample quiz to see if you have mastered the material in this chapter.

SAMPLE QUIZ

Quiz #3

1. Give the symbol for each of the following elements:

    (a) magnesium _____; (b) manganese _____

2. Determine the number of atoms of each element; write the name of the element and the total number of atoms in each of the following molecules:

    (a) $C_2H_6O_4S$ (dimethyl sulfate) _____

    _____

    (b) $C_{15}H_{11}I_4NO_4$ (thyroxine) _____

    _____

3. Classify the following changes as physical or chemical:

    (a) Dissolving sugar in water _____

    (b) Fermentation to produce wine _____

(c) Souring of milk _____

(d) Burning of wood _____

4. Calculate the quantity of heat in kilocalories required to raise the temperature of 165 g of iron from $1200°C$ to $1400°C$. Specific heat of iron = 0.108 cal/g • $°C$.

5. Calculate the volume in liters ($20°C$) of 525 g of a liquid whose specific gravity is

$$1.10^{20°/4}.$$

Solutions and Answers for Quiz #3

1. (a) Mg; (b) Mn

2. (a) 2 atoms carbon, 6 atoms hydrogen, 4 atoms oxygen, 1 atom sulfur; 13 atoms total

   (b) 15 atoms carbon, 11 atoms hydrogen, 4 atoms iodine, 1 atom nitrogen, 4 atoms oxygen; 35 atoms total

3. (a) physical; (b) chemical; (c) chemical; (d) chemical

4. $0.108 \frac{cal}{g • °C} \times 165 \ g \times (1400 - 1200)°C \times \frac{1 \ kcal}{1000 \ cal} = 3.56 \ kcal$

5. $1.10 \times 1.00 \frac{g}{mL} = 1.10 \frac{g}{mL}$

   $525 \ g \times \frac{1 \ mL}{1.10 \ g} \times \frac{1 \ L}{1000 \ mL} = 0.477 \ L$

Review Exam #1 (Chapters 1 to 3) [Answers are in ( ) to the right of each question.]

1. A study of the breakdown of glucose (a sugar) in the human body would be classified as:

    A. organic chemistry   (E)

    B. inorganic chemistry

    C. analytical chemistry

    D. physical chemistry

    E. biochemistry

2. How many significant digits are in the number 0.0707?

    A. One   (C)

    B. Two

    C. Three

    D. Four

    E. Five

3. The number 12.750 rounded off to three significant digits is:

    A. 12.7   (B)

    B. 12.8

    C. 13.0

    D. 13.7

    E. 12.0

4. Ag is the chemical symbol for the element:

    A. gold   (D)

    B. iron

    C. aluminum

    D. silver

    E. mercury

5. The chemical symbol for the element iron is:

    A. I   (D)

    B. In

    C. Rn

    D. Fe

    E. F

6. Which one of the following is <u>not</u> an example of a chemical change?

    A. The rusting of iron   (C)

    B. The tarnishing of silver

    C. The boiling of water

    D. The burning of a piece of paper

    E. The souring of milk

7. Which one of the following would <u>not</u> be classified as a physical property of sodium chloride?

    A. Colorless solid   (D)

    B. Density - 2.17 g/mL

    C. Melting point = $804^{O}C$

    D. Decomposes in water solution under a direct electric current to produce chlorine, hydrogen and sodium hydroxide

    E. 1.0 g dissolves in 2.8 mL water at $25^{O}C$

8. The prefix meaning 1000 units is:

    A. kilo-   (A)

    B. deci-

    C. centi-

    D. milli-

    E. none of these

9. The ratio of the density of one substance to the density of some standard such as water at 4°C is the definition of:

   A. specific heat   (B)

   B. specific gravity

   C. a calorie

   D. density

   E. none of these

10. The energy an object possesses due to its relative position in space or composition is:

    A. radiant energy   (B)

    B. potential energy

    C. chemical energy

    D. kinetic energy

    E. electrical energy

11. Mercury(I) chloride (calomel) consists of 2 atoms of mercury and 2 atoms of chlorine. The formula for one unit of mercury(I) chloride is:

    A. Me_2Cl_2 (E)

    B. My_2Cl_2

    C. Mg_2Cl_2

    D. Hg_2C_2

    E. Hg_2Cl_2

12. Perform the following operations and express your answer in scientific notation to three significant digits:

    $5.20 \times 10^{-3} + 1.2 \times 10^{-4}$

    A. 6.40×10^{-3} (E)

    B. 6.40×10^{-4}

    C. 6.40

    D. 5.32

    E. 5.32×10^{-3}

38

13. Convert 68°F to °C.

    A. 4°C   (C)

    B. 36°C

    C. $2\overline{0}$°C

    D. 154°C

    E. 124°C

14. Convert $-1\overline{0}$°C to °F.

    A. -12°F   (D)

    B. 23°F

    C. $5\overline{0}$°F

    D. 14°F

    E. -14°F

15. Convert $9\overline{00}$ K to °C.

    A. 673°C   (B)

    B. 627°C

    C. 527°C

    D. 1193°C

    E. 981°C

16. Convert 65 mg to kg.

    A. 6.5×10^{-5} kg   (A)

    B. 6.5×10^{-6} kg

    C. 6.5 kg

    D. 6.5×10^{3}

    E. 6.5×10^{6}

17. Convert 0.41 mg to cg.

    A. 4.1×10^{-2} cg   (A)

    B. 4.1×10^{-3} cg

    C. 4.1 cg

    D. 41 cg

    E. 4.1×10^{2} cg

18. Add the following masses and express the answer in grams:

    3.000000 kg, 45$\overline{0}$ mg, 30.0 cg, 0.21 dg

    A. 3000.690 g   (B)

    B. 3000.771 g

    C. 3001.160 g

    D. 3000.501 g

    E. 3690.0 g

19. Calculate the density of an object that has a mass of 20.0 g and occupies a volume of 5.00 mL.

    A. 0.200 g/mL   (B)

    B. 4.00 g/mL

    C. 20.0 g/mL

    D. 5.00 g/mL

    E. Insufficient data given

20. The density of concentrated sulfuric acid is 1.84 g/mL. What volume in liters would have a mass of 75.0 g?

    A. 138 L   (C)

    B. 0.138 L

    C. 4.08×10^{-2} L

    D. 4.08 L

    E. 40.8 L

21. The specific gravity of ethyl ether is $0.708^{25°}/4$.
Calculate the number of milliliters in 0.0350 kg of ethyl ether.

A. 2.48×10^{-2} mL   (B)

B. 49.4 mL

C. 4.94×10^{-2} mL

D. 24.8 mL

E. 2.48×10^{2} mL

22. The density of an organic liquid is 1.20 g/mL. Calculate the number of liters in 400 g of the liquid.

A. 480 L   (B)

B. 0.333 L

C. 0.120 L

D. 0.480 L

E. 1,200 L

23. Calculate the number of calories required to raise the temperature of 200 g of water from $20.0°C$ to $40.0°C$. The specific heat of water is 1.00 cal/g $\cdot$ $°C$.

A. 4000 cal   (A)

B. 10.0 cal

C. 400 cal

D. 2000 cal

E. 1000 cal

24. Calculate the number of kilocalories required to raise the temperature of 325 g of aluminum from $25.0°C$ to $50.0°C$. The specific heat of aluminum is 0.217 cal/g $\cdot$ $°C$.

A. 1760 kcal   (C)

B. 8.12 kcal

C. 1.76 kcal

D. 59.9 kcal

E. 70.5 kcal

25. Calculate the number of grams of aluminum used if 85.0 cal of heat is absorbed when aluminum is heated from 30.0°C to 60.0°C. The specific heat of aluminum is 0.217 cal/g · °C.

A. 553 g   (E)

B. 1.63 g

C. 1.18 x 10^3 g

D. 2.83 g

E. 13.1 g

CHAPTER 4

THE STRUCTURE OF THE ATOM

In Chapter 3 we mentioned atoms; in this chapter we will consider the atom in more detail. We will consider the various particles that make up the atom and the arrangement of these particles in the atom.

SELECTED TOPICS

1. The mass of an atom is extremely small. For example, the mass of a carbon atom is estimated to be 2.00×10^{-23} g. It would be very impractical to use such small numbers in the calculations chemists must perform; therefore, a scale of <u>atomic mass units</u> of the elements has been devised. Carbon-12 (an isotope of carbon, see Selected Topic #4) is the reference element and is assigned a value of exactly 12 atomic mass units (amu). Therefore, one amu is defined as 1/12 of the mass of one carbon-12 atom. The atomic mass of other elements is based on carbon-12. The relative percent of the various isotopes occurring in nature of each element determines the atomic mass of the element.

    The atomic mass scale is analogous to the derivation of the inch scale in linear measure. The "foot" as a linear unit of measure was defined many years ago by the British as equivalent to the length of the foot of one of their kings, and the inch was defined a 1/12 the length of one foot. Thus, the inch scale of linear measurement is related to the foot just as the atomic mass unit is related to the actual mass of the carbon-12 atom. See the following tables.

Mass of an Atom in amu		Mass of Atom in Relation to C-12
1.00 amu	is equal to	1/12 of mass of C-12 atom
4.00 amu	is equal to	1/3 of mass of C-12 atom
16.00 amu	is equal to	4/3 of mass of C-12 atom
24.00 amu	is equal to	2 times mass of C-12 atom
144.00 amu	is equal to	12 times mass of C-12 atom

Linear Measurement in Inches		Linear Measurement in Feet
1 inch	is equal to	1/12 of one foot
4 inches	is equal to	1/3 of one foot
16 inches	is equal to	4/3 of one foot
24 inches	is equal to	2 feet
144 inches	is equal to	12 feet

See Section 4-1 in your text.

2. <u>Atoms</u> consist essentially of three fundamental particles: <u>protons</u> (p), <u>neutrons</u> (n), and <u>electrons</u> (e-). Protons are positively charged particles. Neutrons are particles which have no charge. Electrons are negatively charge particles. Table 4-1 in your text summarizes these three particles. You must know the approximate mass (amu) and relative charge for each of these particles. See Section 4-3 in your text.

3. The protons and neutrons are located in the nucleus of the atom. The electrons are located outside the nucleus. The <u>atomic number</u> of an element is equal to the number of protons in the nucleus. The mass of an atom, for all practical purposes, refers to the mass of the nucleus, since the nucleus contains the protons and neutrons. The mass of an electron can be discounted because it is so small (1/1838 of the mass of a hydrogen atom). The <u>mass number</u> of an element is determined by adding the number of protons and the number of neutrons, each of which has a mass of essentially 1 amu.

mass number = number of protons + number of neutrons

number of neutrons = mass number - atomic number

An atom of an element may be represented by the symbol $^{A}_{Z}E$, where E represents the symbol of the element, A represents the mass number, and Z represents the atomic number. Thus in $^{23}_{11}Na$, the element sodium (Na) has a mass number of 23 and an atomic number of 11. See Section 4-4 in your text.

4. <u>Isotopes</u> are atoms of an element that have a different mass number but the same atomic number. Since the atomic number remains the same, isotopes have the same number of protons and electrons, but different numbers of neutrons. See Section 4-5 in your text.

5. Electrons exist in <u>principal energy levels</u> outside the nucleus of an atom. The electrons in levels close to the nucleus have lower energy than do electrons in levels further from the nucleus. The electrons in the various levels are designated by the whole numbers 1, 2, 3, 4, 5, 6, and 7. The first prinicipal energy level (n = 1) is the most stable and is located closest to the nucleus. Each succeeding energy level is further away from the nucleus. The maximum number of electrons at principal energy levels is found by the formula $2n^2$, where n = the number of the principal energy level, integers of 1 to 7. See Section 4-6 in your text.

6. <u>Valence electrons</u> are those electrons which are located in the highest principal energy level of the atom. Electron-dot formulas are also known as Lewis structures. In these structures symbols of the elements are used with electrons represented by dots. Only the outermost electrons, that is, the valence electrons, are actually shown. See Section 4-7 in your text.

7. Principal energy levels are further subdivided into <u>sublevels</u>, designated <u>s</u>, <u>p</u>, <u>d</u>, and <u>f</u>. Each of these sublevels is arranged in order from lowest energy to highest energy. The lower energy sublevels are filled first, and then the higher energy sublevels. The order of increasing energy levels is as follows: 1<u>s</u><2<u>s</u><2<u>p</u><3<u>s</u><3<u>p</u><4<u>s</u><3<u>d</u><4<u>p</u><5<u>s</u><4<u>d</u><5<u>p</u><6<u>s</u><(4<u>f</u><5<u>d</u>)<6<u>p</u><7<u>s</u><(5<u>f</u><6<u>d</u>). The maximum numbers of electrons in the <u>s</u>, <u>p</u>, <u>d</u>, and <u>f</u> sublevels are 2, 6, 10, and 14 respectively. (Note: 4 is added each time.)

Figure 4-5 in your text and the following diagram give a simplified way of remembering the order of filling the sublevels:

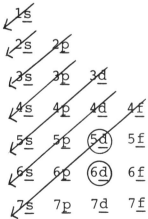

In drawing this diagram, use the following procedure:

1. On separate lines, write the principal energy levels with their sublevels to the _f_ sublevel. (The number of the principal energy level equals the number of the sublevels, but we need not go beyond the _f_ sublevel as elements with electrons beyond the _f_ sublevel have yet to be discovered.)

2. Draw diagonal lines for the order of filling these sublevels. Stop with the 6_d_ sublevel since no elements at present have been discovered that have electron configurations beyond the 6_d_ sublevel.

3. Circle the 5_d_ and 6_d_ sublevels because only one electron is placed in each of these sublevels initially. After filling the 6_s_, skip the 4_f_ and place one electron in the 5_d_, then go back to the 4_f_ and fill it. Once the 4_f_ is filled, return to the 5_d_ and completely fill it to its maximum of 10 electrons. The process repeats itself for the 6_d_ and 5_f_ sublevels. After filling the 7_s_, skip the 5_f_ and place one electron in the 6_d_, then go back to the 5_f_ and fill it. Once the 5_f_ is filled return to the 6_d_. (Figure 4-7 in your text is another diagram for remembering the order of filling the sublevels.)

See Section 4-8 in your text.

PROBLEM 4-1

For each of the following atoms, calculate the number of protons and neutrons in the nucleus and the number of electrons outside the nucleus.

(a) $^{14}_{7}N$ (a) _____

(b) $^{40}_{20}Ca$ (b) _____

(c) $^{39}_{19}K$ (c) _____

(d) $^{238}_{92}U$ (d) _____

ANSWERS TO PROBLEM 4-1

	Nucleus	Outside Nucleus
(a)	7p 7n	7e-

(b)  20e-

(c) 19e-

(d) 92e-

If your answers are correct, proceed to Problem 4-2. If your answers are incorrect, review Section 4-4 (Example 4-1 and Study Exercise 4-1) in your text.

Proceed to Problem 4-2.

PROBLEM 4-2

Calculate the atomic mass to four significant digits for copper given the following data:

Isotope	Exact Atomic Mass (amu)	% Abundance in Nature
^{63}Cu	62.9298	69.09
^{65}Cu	64.9278	30.91

ANSWER AND SOLUTION TO PROBLEM 4-2

63.55 amu

$$(62.9298 \text{ amu})(0.6909) + (64.9278 \text{ amu})(0.3091) = 63.55 \text{ amu}$$

Did you convert the percents (69.09 and 30.91) to decimal form by dividing by 100?

If your answer is correct, proceed to Problem 4-3. If your answer is incorrect, review Section 4-5 (Examples 4-2 and 4-3 and Study Exercises 4-2 and 4-3) in your text.

Proceed to Problem 4-3.

PROBLEM 4-3

Calculate the maximum number of electrons that may occupy the following principal energy levels:

(a)   3

(b)   4

(c)   5

(a) _____

(b) _____

(c) _____

ANSWERS AND SOLUTIONS TO PROBLEM 4-3

The maximum number of electrons in the principal energy level follows the formula $2n^2$.

(a)   18e- $2(3)^2 = 2 \times 9 = 18$

(b)   32e- $2(4)^2 = 2 \times 16 = 32$

(c)   50e- $2(5)^2 = 2 \times 25 = 50$

If your answers are correct, proceed to Problem 4-4.   If your answers are incorrect, review Section 4-6 (Example 4-4 and Study Exercise 4-4) in your text.

Proceed to Problem 4-4.

PROBLEM 4-4

Diagram the atomic structure for each of the following atoms. Indicate the number of protons and neutrons, and arrange the electrons in principal energy levels.

(a) $^{12}_{6}C$

(a) _____

(b) $^{24}_{12}Mg$

(b) _____

(c) $^{25}_{12}Mg$

(c) _____

(d) $^{32}_{16}S$

(d) _____

(e) $^{27}_{13}Al$

(e) _____

(f) $^{35}_{17}Cl$ (f) _____

ANSWERS TO PROBLEM 4-4

		(1)	(2)	(3) Principal Energy Levels
(a)	6p 6n	2e-	4e-	
(b)	12p 12n	2e-	8e-	2e-
(c)	12p 13n	2e-	8e-	2e-
(d)	16p 16n	2e-	8e-	6e-
(e)	13p 14n	2e-	8e-	3e-
(f)	17p 18n	2e-	8e-	7e-

If your answers are correct, proceed to Problem 4-5. If your
answers are incorrect, review Section 4-6 (Study Exercise 4-5)
in your text.

Proceed to Problem 4-5.

PROBLEM 4-5

Write the electron-dot formulas for each of the following
atoms:

(a) $^{40}_{18}Ar$ (a) _____

(b) $^{12}_{6}C$ (b) _____

(c) $^{34}_{16}S$ (c) _____

(d) $^{14}_{7}N$ (d) _____

(e) $^{16}_{8}O$ (e) _____

(f) $^{19}_{9}F$ (f) _____

49

ANSWERS TO PROBLEM 4-5

(a)    :Är: Remember that in the electron-dot configura-
              tions, only valence electrons are shown. How
              do you determine which electrons are the va-
              lence electrons?

(b)    ·C·

(c)   ·S:

(d)    ·N:

(e)    ·O:

(f)   :F:

If your answers are correct, proceed to Problem 4-7. If your
answers are incorrect, review Section 4-7 (Study Exercise 4-6)
in your text and then try a similar problem, Problem 4-6.

Proceed to Problem 4-6.

PROBLEM 4-6

Write the electron-dot formulas for each of the following
atoms:

(a) $^{20}_{10}Ne$ (a) _____

(b) $^{27}_{13}Al$ (b) _____

(c) $^{11}_{5}B$ (c) _____

(d) $^{31}_{15}P$ (d) _____

(e) $^{35}_{17}Cl$ (e) _____

(f) $^{9}_{4}Be$ (f) _____

ANSWERS TO PROBLEM 4-6

(a)    :Ne:

(b)   Al

(c)   B

(d)   P

(e)   Cl

(f)   Be

If your answers are correct, proceed to Problem 4-7. If your answers are incorrect, then re-read Section 4-7 in your text. Re-work Problem 4-5 and 4-6.

Proceed to Problem 4-7.

PROBLEM 4-7

(1) Write the electronic configuration in sublevels for each of the following atoms. (2) Give the number of valence electrons for each.

(a) $^{1}_{1}H$ (a) _____

(b) $^{14}_{7}N$ (b) _____

(c) $^{19}_{9}F$ (c) _____

(d) $^{16}_{8}O$ (d) _____

(e) $^{20}_{10}Ne$ (e) _____

(f) $^{64}_{30}Zn$ (f) _____

                                        _____

ANSWERS TO PROBLEM 4-7

(a) $1\underline{s}^{1}$ (1)

(b) $1\underline{s}^{2}, 2\underline{s}^{2} 2\underline{p}^{3}$ (5)

(c) $1\underline{s}^{2}, 2\underline{s}^{2} 2\underline{p}^{5}$ (7)

(d) $1\underline{s}^2$, $2\underline{s}^2$ $2\underline{p}^4$ (6)

(e) $1\underline{s}^2$, $2\underline{s}^2$ $2\underline{p}^6$ (8)

(f) $1\underline{s}^2$, $2\underline{s}^2$ $2\underline{p}^6$, $3\underline{s}^2$ $3\underline{p}^6$ $3\underline{d}^{10}$, $4\underline{s}^2$
     The 3$\underline{d}$ sublevel fills after the 4$\underline{s}$.

   Or $1\underline{s}^2$, $2\underline{s}^2$ $2\underline{p}^6$, $3\underline{s}^2$ $3\underline{p}^6$, $4\underline{s}^2$, $3\underline{d}^{10}$

   (2, the valence electrons are the electrons in the highest
   principal energy level, that is the 4$\underline{s}$)

If your answers are correct, you have completed this chapter.
If your answers are incorrect, review Section 4-8 in your text
and then try a similar problem, Problem 4-8.

Proceed to Problem 4-8.

PROBLEM 4-8

(1) Write the electronic configuration in sublevels for each of
the following atoms.   (2) Give the number of valence electrons
for each.

(a) $^{27}_{13}$Al   (a) _____

(b) $^{32}_{16}$S   (b) _____

(c) $^{40}_{18}$Ar   (c) _____

(d) $^{58}_{28}$Ni   (d) _____

ANSWERS TO PROBLEM 4-8

(a) $1\underline{s}^2$, $2\underline{s}^2$ $2\underline{p}^6$, $3\underline{s}^2$ $3\underline{p}^1$ (3)

(b) $1\underline{s}^2$, $2\underline{s}^2$ $2\underline{p}^6$, $3\underline{s}^2$ $3\underline{p}^4$ (6)

(c) $1\underline{s}^2$, $2\underline{s}^2$ $2\underline{p}^6$, $3\underline{s}^2$ $3\underline{p}^6$ (8)

(d) $1\underline{s}^2$, $2\underline{s}^2$ $2\underline{p}^6$, $3\underline{s}^2$ $3\underline{p}^6$ $3\underline{d}^8$, $4\underline{s}^2$

     The 3$\underline{d}$ sublevel fills after the 4$\underline{s}$.

     Or $1\underline{s}^2$, $2\underline{s}^2$ $2\underline{p}^6$, $3\underline{s}^2$ $3\underline{p}^6$, $4\underline{s}^2$, $3\underline{d}^8$

     (2, the valence electrons are the electrons in the highest
     principal energy level, that is the 4$\underline{s}$)

If your answers are correct, you have completed this chapter.
If your answers are incorrect, re-read Section 4-8 in your
text. Re-work Problems 4-7 and 4-8.

These problems conclude the chapter on the structure of the
atom.

Now take the sample quiz to see if you have mastered the
material in this chapter.

SAMPLE QUIZ

<u>Quiz #4</u>

1. For each of the following atoms, draw the nucleus and
   label the number of protons and neutrons, and arrange the
   electrons in principal energy levels.

   (a) $^{31}_{15}P$

   (b) $^{38}_{18}Ar$

2. Calculate the maximum number of electrons that may occupy
   the following principal energy levels:

   (a) 6 _____   (b) 7 _____

3. (1) Write the electronic configuration in <u>sublevels</u> for the
   following atoms. (2) Give the number of valence electrons
   of each.

   (a) $^{24}_{12}Mg$ _____

   (b) $^{51}_{23}V$ _____

4. Write the electron-dot formulas for the following atoms:

   (a) $^{14}_{7}N$ _____   (b) $^{35}_{17}Cl$ _____

5. Calculate the atomic mass to four significant digits for
   the hypothetical element X, given the following data:

Element	Exact Atomic Mass (amu)	% Abundance in Nature
^{10}X	10.00	80.00
^{12}X	12.10	20.00

Solutions and Answers for Quiz #4

1. (a)   (15p 16n)   2e-   8e-   5e-

   (b)   (18p 20n)   2e-   8e-   8e-

                        (1)   (2)   (3)

2. (a) $2 \times 6^2 = 72$ (b) $2 \times 7^2 = 98$

3. (a) $1\underline{s}^2, 2\underline{s}^2 2\underline{p}^6, 3\underline{s}^2$ (2)

   (b) $1\underline{s}^2, 2\underline{s}^2 2\underline{p}^6, 3\underline{s}^2 3\underline{p}^6 3\underline{d}^3, 4\underline{s}^2$

        or $1\underline{s}^2, 2\underline{s}^2 2\underline{p}^6, 3\underline{s}^2 3\underline{p}^6, 4\underline{s}^2, 3\underline{d}^3$ (2)

4. (a) $\cdot \overset{\bullet}{\underset{\bullet}{N}} \colon$ (b) $\colon \overset{\bullet}{\underset{\bullet\bullet}{Cl}} \colon$

5. (10.00 amu x 0.8000) + (12.10 amu x 0.2000) = 10.42 amu

CHAPTER 5

THE PERIODIC CLASSIFICATION OF THE ELEMENTS

In previous chapters of your text (Sections 3-7 and 4-8), we mentioned the periodic table in regard to the separation of metals from nonmetals and the filling of sublevels. In this chapter we will consider the classification of the elements in the periodic table and some general characteristics of groups of elements.

SELECTED TOPICS (In reading this section, refer to the periodic table on the inside front cover of your text.)

1. The periodic table is the arrangement of the elements in a tabular form consisting of horizontal rows and vertical columns. The periodic table is based on the periodic law. The periodic law states that the appearance of elements with similar chemical properties at regular intervals in the periodic table occurs because they are listed in order of increasing atomic number. In the periodic table, elements with similar electronic configurations are grouped together. Various chemical properties of an element can then be predicted from its location in the periodic table. See Section 5-1 in your text.

2. The periodic table is arranged in eighteen vertical columns called groups and seven horizontal rows called periods. Elements in the same group have similar electronic configuration and thus have similar chemical properties. Most of the eighteen groups or families are indicated with Roman numerals and an A or B, or Arabic numerals. These Arabic numerals are written above the Roman numerals in the periodic table and in parenthesis next to the Roman numeral in your text and this book. The A group elements (1 and 2 and 13 to 18) are called the representative elements. The B group elements (3 to 7, 11 and 12) and group VIII (three vertical columns in this group, 8, 9, 10) are called the transition elements. A period begins with group IA (1) and ends with group VIIIA (18). Special names are given to some of the groups of elements. For example group VIIIA (18) is called the noble gases. See Section 5-2 in your text.

3. Because each group of elements exhibits similar chemical properties, we can use the periodic table to predict the general characteristics and estimate the properties of elements. From the position of an element in the periodic table, we can determine whether the element is a metal, nonmetal, or metalloid (semimetal). Elements to the right of the colored stair step line (see the periodic table in your text) are nonmetals; those to the left are metals; and those that lie on the colored stair step line are called metalloids (semimetals, not aluminum - a metal). Metalloids have both metallic and nonmetallic properties (see

Section 3-7). See Section 5-3, number 1, in your text.

4. Valence electrons are the electrons in the highest principal energy level. In the A group elements, the Roman group numeral or the <u>unit</u> digit of the Arabic numeral gives the number of valence electrons. For group VIIIA (18) the number of valence electrons is 8, except for helium which is 2. For the transition elements [B group elements and group VIII (8, 9, 10) elements] this generalization does not hold, since the transition elements usually have 1 or 2 valence electrons. See Section 5-3, number 2, in your text.

5. Elements in the same group have similar chemical properties and similar electronic configurations. This is analogous to members of a family looking alike! See Section 5-3, number 3, in your text.

6. In the A group elements, the metallic properties increase within a given group with increasing atomic numbers, and the nonmetallic properties decrease. In moving from left to right in a given period in the periodic table, the A group elements become more nonmetallic. As we mentioned previously, the nonmetals are located to the right of the colored stair step line. The following diagram summarizes these generalizations:

```
                    Nonmetallic properties increase
                    ----------------------------------->
     Metallic       |
     properties     |
     increase       ↓
```

See Section 5-3, number 4, in your text.

7. There is a gradual change of many physical and chemical properties within a given group with increasing atomic number. One of these properties is atomic radius. In group IIA (2, the alkaline earth metals), the radii are as

follows: Be = 125 pm (pm, picometer, 1 pm = 10^{-12} m, see Table 2-1 in your text), Mg = 145 pm, Ca = 174 pm, Sr = 191 pm, and Ba = 198 pm. The increase in atomic radii with an increase in atomic number within a given group is due to the presence of electrons in higher principal energy levels as we move down the group to the next period. See Section 5-3, number 5, in your text.

IF IN SOME OF THE FOLLOWING PROBLEMS, IF YOU ARE NOT FAMILIAR WITH THE SYMBOLS FOR THE ELEMENTS, LOOK THEM UP INSIDE THE FRONT COVER OF YOUR TEXT.

PROBLEM 5-1

Using the periodic table, classify the following elements as metals, nonmetals, or metalloids (semimetals).

(a) sodium   (a) _____

(b) nickel   (b) _____

(c) arsenic   (c) _____

(d) sulfur   (d) _____

ANSWERS TO PROBLEM 5-1

(a) metal   To the left of the colored stair step line - metal.

(b) metal   Same as (a) - metal.

(c) metalloid   On the colored stair step line - metalloid
     (semimetal)   (semimetal).

(d) nonmetal   To the right of the colored stair step line - nonmetal.

If your answers are correct, proceed to Problem 5-2. If your answers are incorrect, review Section 5-3, number 1 (Study Exercise 5-1) in your text.

Proceed to Problem 5-2.

PROBLEM 5-2

Using the periodic table, indicate the number of valence electrons for the following elements:

(a) phosphorus   (a) _____

(b) potassium   (b) _____

(c) bromine   (c) _____

(d) argon   (d) _____

ANSWERS TO PROBLEM 5-2

(a) 5   Group VA (1$\underline{5}$) - 5 valence electrons.

(b) 1   Group IA ($\underline{1}$) - 1 valence electron.

(c) 7   Group VIIA (1$\underline{7}$) - 7 valence electrons.

(d) 8   Group VIIIA (1$\underline{8}$), not helium - 8 valence electrons.

If your answers are correct, proceed to Problem 5-3. If your answers are incorrect, review Section 5-3, number 2 (Study Exercise 5-2) in your text.

Proceed to Problem 5-3.

PROBLEM 5-3

Given are the electronic configurations for four elements. Pair the elements you would expect to show similar chemical properties.

(a) $1\underline{s}^2$, $2\underline{s}^2$ $2\underline{p}^6$, $3\underline{s}^2$ $3\underline{p}^1$

(b) $1\underline{s}^2$, $2\underline{s}^2$ $2\underline{p}^6$

(c) $1\underline{s}^2$, $2\underline{s}^2$ $2\underline{p}^6$, $3\underline{s}^2$ $3\underline{p}^6$

(d) $1\underline{s}^2$, $2\underline{s}^2$ $2\underline{p}^6$, $3\underline{s}^2$ $3\underline{p}^6$ $3\underline{d}^{10}$, $4\underline{s}^2$ $4\underline{p}^1$ _____

ANSWERS TO PROBLEM 5-3

(a) and (d)   The outermost electronic configuration in

both cases is an $\underline{s}^2\underline{p}^1$.

(b) and (c)   The outermost electronic configuration in

both cases is an $\underline{s}^2\underline{p}^6$.

If your answers are correct, proceed to Problem 5-4. If your answers are incorrect, review Section 5-3, number 3 (Study Exercises 5-3 and 5-4) in your text.

Proceed to Problem 5-4.

PROBLEM 5-4

Using the periodic table, indicate which element in each of the following pairs of elements is the more metallic:

(a)   sulfur and selenium   (a) _____

(b)   phosphorus and aluminum   (b) _____

(c)   silicon and lead   (c) _____

(d)   sodium and potassium   (d) _____

ANSWERS TO PROBLEM 5-4

(a)   selenium   Selenium (Se) is below sulfur (S) in the same group and hence is more metallic.

(b)   aluminum   Aluminum (Al) is to the left of phosphorus (P) in the same period and hence is more metallic.

(c)   lead   Lead (Pb) is below silicon (Si) in the same group and hence is more metallic.

(d)   potassium   Potassium (K) is below sodium (Na) in the same group and hence is more metallic.

If your answers are correct, proceed to Problem 5-5. If your answers are incorrect, review Section 5-3 number 4 (Study Exercise 5-5) in your text.

Proceed to Problem 5-5.

PROBLEM 5-5

Using the periodic table, indicate which element in each of the following pairs of elements has the greater atomic radius:

(a)   sulfur and selenium   (a) _____

(b)   gold and copper   (b) _____

(c)   lead and silicon   (c) _____

(d)   arsenic and bismuth   (d) _____

ANSWERS TO PROBLEM 5-5

(a)   selenium   Selenium (Se) is below sulfur (S) in the same group, hence selenium has additional electrons in a higher principal energy level and a greater atomic radius.

(b)   gold   For (b), (c), and (d) - same reasoning as above.

(c)   lead

(d)   bismuth

If your answers are correct, you have completed this chapter. If your answers are incorrect, review Section 5-3, number 5 (Study Exercise 5-6) in your text.

These problems conclude the chapter on the periodic classification of the elements.

Now take the sample quiz to see if you have mastered the material in this chapter.

SAMPLE QUIZ

Quiz #5 - You may use the periodic table and the list of symbols of the elements.

1. Classify the following elements as metals, nonmetals, or metalloids (semimetals).

   (a) potassium _____   (b) titanium _____

   (c) antimony _____   (d) sulfur _____

2. Indicate the number of valence electrons for the following elements:

   (a) calcium _____   (b) selenium _____

   (c) indium _____   (d) krypton _____

3. Given the electronic configuration for four elements. Pair the elements you would expect to show similar chemical properties.

   (a) $1s^2$, $2s^2$ $2p^1$

   (b) $1s^2$, $2s^2$ $2p^6$, $3s^2$ $3p^6$

   (c) $1s^2$, $2s^2$ $2p^6$, $3s^2$ $3p^6$ $3d^{10}$, $4s^2$ $4p^6$

   (d) $1s^2$, $2s^2$ $2p^6$, $3s^2$ $3p^6$ $3d^{10}$, $4s^2$ $4p^1$

4. Indicate which element in each of the following pairs of elements is the more metallic:

   (a) cesium and potassium _____

   (b) aluminum and gallium _____

5. Indicate which element in each of the following pairs of elements has the greater atomic radius:

   (a) bromine and fluorine _____

   (b) copper and gold _____

60

Answers for Quiz #5

1. (a) metal; (b) metal; (c) metalloid (semimetal); (d)
   nonmetal

2. (a) 2; (b) 6; (c) 3; (d) 8

3. (a) and (d); (b) and (c)

4. (a) cesium; (b) gallium

5. (a) bromine; (b) gold

CHAPTER 6

THE STRUCTURE OF COMPOUNDS

In Chapter 4 we considered the structure of the atom. In Chapter 6 we will discuss joining atoms to form compounds and the shape of these compounds. We will also consider the periodic table in more detail and see how it can be used to predict properties of elements.

SELECTED TOPICS (In reading this section, refer to the periodic table on the inside cover of your text.)

1. Compounds are composed of chemical bonds. Chemical bonds are the attractive forces that hold atoms together to make compounds. There are two general types of bonds between atoms in a compound: (1) ionic bonds and (2) covalent bonds. These bonds are formed from the valence electrons of the atoms in a compound. The basis of this formation is the rule of eight (octet rule) - most atoms attempt to obtain a stable configuration of eight valence electrons around the atom. An exception to the rule of eight is the rule of two - a completed first principal energy level is also a stable configuration. See Section 6-1 in your text.

2. Formation of ionic and covalent bonds depends on (1) the ionization energy of an element - amount of energy required to remove the most loosely bound electron from an atom, ion, or molecule of the element to form a cation (positively charged ion) and (2) the electron affinity of an element - amount of energy given off when an atom or ion picks up an extra electron to form an anion (negatively charged ion). Table 6-1 lists some common metals with the formulas of their cations and names. Table 6-2 list some common nonmetals with the formulas of their anions and names. You must know Tables 6-1 and 6-2. By using the periodic table, this task is simplified. In Table 6-1, the positive ionic charge on those cations marked with an asterisk (*), can be related to the periodic table. In general the group Roman numeral for all the A group elements represents this positive ionic charge. For example, because aluminum is in group IIIA, you would

    expect it to have a 3^+ ionic charge, and that is its positive ionic charge according to Table 6-1. This postive ionic charge is the most common ionic charge for the element. You must memorize the ionic charge for the other cations not marked with an asterisk (*). In the name of the cation, the Roman numeral in parenthesis indicates the ionic charge on each atom in the ion. In Table 6-2, the negative ionic charge on these anions, can be determined by subtracting 8 from the group Roman numeral in the periodic table. For example, because bromine is in group VIIA, you would expect it to have a 1⁻ ionic charge (VII - 8 = -1) and that is its negative ionic charge according to Table 6-2. The hydride ion (H^-) is an exception. Its negative charge can be determined by subtracting 2 from the group Roman numeral (IA), hence I - 2 = -1. The reason for subtracting 2 is that 2 electrons complete the first principal energy level. See Section 6-2 in your text.

3. <u>Oxidation number</u> (or oxidation state) is a positive or negative whole number assigned to an element in a compound or ion. It is based on certain rules (see Section 6-3 in your text). All atoms have an oxidation number of zero when they are uncombined. This oxidation number changes when a reaction occurs and the atom loses or gains electrons. An atom is assigned a positive oxidation number if it loses electrons. An atom is assigned a negative oxidation number if it gains electrons. The oxidation number of a monatomic ion (ion containing a single atom) is the same as its ionic charge. This arbitrary assignment of oxidation numbers is based on the rules given in Section 6-3 in your text.

4. <u>Ionic bonds</u> are formed by the transfer of electrons from one atom to another. The structure of monatomic ions (cations and anions) can be determined by removing or adding electrons to the structure of the corresponding atoms. A positively charged monoatomic cation is formed by removing electrons (electrons are negatively charged) from an atom. The sodium cation (Na^+) is formed from the sodium atom by removing one electron from the neutral atom. The sodium ion then has 11 protons (positive) and only 10 electrons (negative) for a 1⁺ net charge. A negatively charged monoatomic ion is formed by adding electrons to an atom.

The chloride anion (Cl^-) is formed from the chlorine atom by adding one electron to the neutral atom. The chloride ion then has 17 protons (positive) and 18 electrons

(negative) for a 1⁻ net charge. The charge on the ion determines the number of electrons added or removed from the atom. The bond formed between oppositely charged particles is an <u>ionic bond</u>. This bond is formed by the transfer of one or more electrons from one atom to another, and the positive and negative charges attract each other. The smallest unit of an ionic compound is a <u>formula unit</u>. See Section 6-4 in your text.

5. A <u>covalent bond</u> is formed by sharing of electrons between two atoms. The smallest unit of covalent compound is a <u>molecule</u>. A covalent bond can be formed by the equal sharing or unequal sharing of electrons between atoms. If the atoms are the same in a diatomic molecule, we have e-qual sharing of electrons. Diatomic molecules that have e-

qual sharing of electrons are H_2, Cl_2, F_2, Br_2, I_2, O_2, and N_2. Unequal sharing of electrons is found in diatomic molecules where the atoms are not the same. The basis of this unequal sharing of electrons is electronegativity. Electronegativity is the degree to which an atom attracts a pair of covalently bonded electrons to itself. In general, the metals have low electronegativities and the non-metals have high electronegativities. A partial series of decreasing electronegativities is as follows: F>O>Cl, N >Br>I, C, S>P, H>B>Si. Elements high in this partial list that form a compound with an element low on the list are

designated with a δ^- (lower case Greek letter, δ ,

meaning partial). The element in the compound low on the

list (low electronegativity) is designated with a δ^+.
The compound hydrogen bromide (HBr) would be depicted as

follows: $\overset{\delta^+}{H} \overset{\delta^-}{Br}$. This type of covalent bond is often referred to as a polar bond or polar covalent bond. In HBr, one electron was contributed by each atom (H and Br) to form an electron-pair between the two atoms. See Section 6-5 in your text.

6. A bond can also be formed when one atom supplies both electrons of the electron-pair. This type of covalent bond is a coordinate covalent bond. An example is in the ammonium ion,

$$\left[\begin{array}{c} H \\ \overset{\bullet\bullet}{} \\ H : N : H \\ \overset{\bullet\,\bullet}{} \\ H \end{array} \right]^+$$

See Section 6-6 in your text.

7. Lewis structures (electron-dot formulas) of molecules and ions show the covalent bonds among atoms in molecules and ions using the rule of eight (octet rule) and dots (:) to represent bonds. For Lewis structures of compounds or ions that we will consider in this book, all elements in the formula except hydrogen obey the rule of eight; that is, there are eight electrons in their last principal energy level. Hydrogen obeys the rule of two. The noble gases have eight electrons in their highest principal energy level, except helium which has two completing the first principal energy level. The noble gases are relatively inert. In writing Lewis structures of negative ions, add one electron to the formula for each negative charge on the ion. In writing Lewis structures of positive ions, subtract one electron from the formula for each positive charge on the ion. Ions consisting of more than one atom with a net positive or negative charge on the ion are

called <u>polyatomic ions</u>. These polyatomic ions are listed in Table 6-5 of your text. To write structural formulas of compounds and ions, use the Lewis structures of molecules and ions. A <u>structural formula</u> is a formula showing the arrangement of atoms within the molecule or ion, using a dash for each pair of electrons shared between atoms. See Section 6-7 in your text.

8. Lewis structures do not give the three-dimensional shape of molecules or ions. For the shape of these molecules or ions we use the **v**alence **s**hell **e**lectron **p**air **r**epulsion **(VSEPR)** model. The basis of this model is Coulomb's law - attractive forces between a proton and an electron increase as the distance between the two particles decreases. Most molecules or polyatomic ions have a central atom to which other atoms are attached. The shape of the molecule or ion depends on the number of pairs of electrons surrounding the central atom. From Coulomb's law these pairs of electrons (bonding and unshared pairs) will arrange themselves about the central atom in such a way as to maximize the distance among all these electron pairs, that is, they get as far apart from each other as possible. In doing this, the electrons minimize the repulsive energy generated between them. In determining the shape and bond angle of molecules or ions we follow certain guidelines given in Section 6-8 of your text. In using these guidelines the Lewis structures must be drawn first and the total number of electron pairs or group of electrons and unshared pairs of electrons around the central atom must be determined. Using this information and relating it to Table 6-6 in your text, the shape and bond angle of a molecule or ion can be determined. See Section 6-8 in your text.

9. The cations, anions, and polyatomic ions are used to write formulas of compounds. To write the correct formula of a compound, you must know or be given the ionic charge on the ion. The sum of the total positive charges must be equal to the sum of the total negative charges (that is, the compound must not possess a net charge) in a correct formula of a compound. When the charge on the positive ion is not equal to the charge on the negative ion, subscripts must be used to balance the positive charges with the negative charges. See Section 6-9 in your text.

10. You can use the periodic table to learn the ionic charges on the cations (Table 6-1) and anions (Table 6-2). It can also help you to estimate the properties of compounds and to predict formulas of compounds and the type of bonding in compounds.

In general, the Roman group numberal represents the maximum positive oxidation number for the element in the group. For nonmetals, calculate the oxidation number by subtracting 8 from the Roman group number. Using these generalizations and the principles of formula writing, you can predict the formulas of compounds containg two different elements. See Section 6-10 - Oxidation Numbers - in your text.

Noticing the trend in various properties of elements or compounds and the position of an element in the periodic table, you can predict numerical values of properties of elements or compounds. See Section 6-10 - Properties - in your text.

Based on the principle that all elements in a group have similar electronic configurations and similar chemical properties, you can predict the formulas of compounds of elements in that group, given the correct formula of one compound containing an element in that group. See Section 6-10 - Formulas - in your text.

Using the periodic table, you can predict whether a compound is primarily ionic or covalently bonded. The greater the difference in electronegativities of the elements, the greater the ionic character of the compound. Therefore, in binary compounds formed between elements in group IA (1, except hydrogen) or group IIA (2) with elements in group VIIA (17) or group VIA (16, oxygen and sulfur only), ionic compounds result. Both fluorine and oxygen have high electronegativities, and a compound formed with fluorine or oxygen and a metal is an ionic compound. All other combinations of binary compounds are considered to be covalent compounds. For ternary and higher compounds, the combination of any element (except hydrogen) with any polyatomic ion gives an ionic compound. See Section 6-10 - Types of Bonding - Ionic or Covalent - in your text.

PROBLEM 6-1

Calculate the oxidation number of the element indicated in each of the following compounds or ions:

(a) S in H_2SO_3 (a) _____

(b) S in H_2SO_5 (b) _____

(c) S in SO_4^{2-} (c) _____

(d) S in $S_2O_3^{2-}$ (d) _____

66

ANSWERS AND SOLUTIONS TO PROBLEM 6-1

(a) + 4

$$2(+1) + ox. no. S + 3(-2) = 0$$
$$+ 2 + ox. no. S - 6 = 0$$
$$ox. no. S - 4 = 0$$
$$ox. no. S = + 4$$

(b) + 8

$$2(+1) + ox. no. S + 5(-2) = 0$$
$$+ 2 + ox. no. S - 10 = 0$$
$$ox. no. S - 8 = 0$$
$$ox. no. S = + 8$$

(c) + 6

$$ox. no. S + 4(-2) = - 2$$
$$ox. no. S - 8 = - 2$$
$$ox. no. S = 8 - 2$$
$$ox. no. S = + 6$$

(d) + 2

$$2(ox. no. S) + 3(-2) = - 2$$
$$2(ox. no. S) - 6 = - 2$$
$$2(ox. no. S) = 6 - 2$$
$$2(ox. no. S) = 4$$
$$ox. no. S = 4/2$$
$$ox. no. S = + 2$$

If you answers are correct, proceed to Problem 6-2. If your answers are incorrect, review Section 6-3 (Example 6-1 and Study Exercise 6-1) in your text.

Proceed to Problem 6-2.

PROBLEM 6-2

Diagram the ionic structure for each of the following ions, indicating the number of protons and neutrons in the nucleus and arranging the electrons in principal energy levels.

(a) $^{9}_{4}Be^{2+}$ (a) _____

(b) $^{27}_{13}Al^{3+}$ (b) _____

(c) $^{14}_{7}N^{3-}$ (c) _____

(d) $^{32}_{16}S^{2-}$ (d) _____

ANSWERS TO PROBLEM 6-2

(a)   (4p
             5n)   2e-   2+

(b)   (13p
             14n)   2e-   8e-   3+

(c) $\overset{7p}{\underset{7n}{\bigcirc}}$ 2e-   8e-   3-

(d) $\overset{16p}{\underset{18n}{\bigcirc}}$ 2e-   8e-   8e-   2-

                    (1)   (2)   (3) Principal Energy Levels

If your answers are correct, proceed to Problem 6-3. If your answers are incorrect, review Section 6-4 (Study Exercise 6-2) in your text.

Proceed to Problem 6-3.

PROBLEM 6-3

Write the electronic configuration in sublevels for the following ions:

(a) $^{24}_{12}Mg^{2+}$

                               (a) _____

(b) $^{31}_{15}P^{3-}$

                               (b) _____

ANSWERS TO PROBLEM 6-3

(a) $1\underline{s}^2, 2\underline{s}^2\ 2\underline{p}^6$

(b) $1\underline{s}^2, 2\underline{s}^2\ 2\underline{p}^6, 3\underline{s}^2\ 3\underline{p}^6$

If your answers are correct, proceed to Problem 6-4. If your answers are incorrect, review Section 6-4 (Study Exercise 6-3) in your text.

Proceed to Problem 6-4.

PROBLEM 6-4

Place a δ^+ above the atom or atoms that are relatively positive and δ^- above the atom or atoms that are relatively negative in the following covalent bonded molecules (see Figure 6-9 in your text).

(a)   HI

                               (a) _____

(b) NF_3

                               (b) _____

(c)   HCl

                               (c) _____

(d) PCl_3 (d) _____

ANSWERS TO PROBLEM 6-4

(a) $H^{\delta+}I^{\delta-}$

(b) $N^{\delta+}F_3^{\delta-}$

(c) $H^{\delta+}Cl^{\delta-}$

(d) $P^{\delta+}Cl_3^{\delta-}$

If your answers are correct, proceed to Problem 6-5. If your answers are incorrect, review Section 6-5 (Study Exercises 6-4 and 6-5) in your text.

Proceed to Problem 6-5.

PROBLEM 6-5

Write Lewis structures and structural formulas for the following molecules or polyatomic ions:

(a) Cl_2 (a) _____

(b) CCl_4 (b) _____

(c) CS_2 (c) _____

(d) ClO_4^- (d) _____

ANSWERS AND SOLUTIONS TO PROBLEM 6-5

	Electron-dot formulas of elements	Electron-dot formulas of molecules or ions	Structural formulas
(a)	$\overset{\times\times}{\underset{\times\times}{\times}}\overset{}{Cl}\overset{}{\wedge}$ $\cdot\,Cl\,\overset{\circ\circ}{\underset{\circ\circ}{\circ}}$	$\overset{\times\times}{\underset{\times\times}{\times}}Cl\overset{\times}{\wedge}\,Cl\,\overset{\circ\circ}{\underset{\circ\circ}{\bullet}}$	Cl——Cl
	Cl has 7 valence electrons.	By sharing an electron from each Cl atom, each Cl atom will have a completed valence energy level of 8.	Draw a dash for each pair of shared electrons.

69

(b) $\overset{\times}{\underset{\times}{\times}}\overset{\times}{C}\times$ $\cdot\overset{\circ\circ}{\underset{\circ\circ}{Cl}}\overset{\circ}{\circ}$ $\overset{\circ\circ}{\cdot Cl}\overset{\circ\circ}{\cdot}\overset{\circ\circ}{\cdot Cl}\cdot\overset{\circ\circ}{\underset{\circ\circ}{Cl}}\overset{\circ}{\circ}$ $\overset{\circ\circ}{\underset{\circ\circ}{Cl}}\overset{\circ}{\circ}$

C has 4 valence
electrons and Cl
has 7.

$\overset{\times}{\underset{\circ\circ}{\circ Cl}}\overset{\times}{\underset{\times}{C}}\overset{\times}{\underset{\circ\circ}{Cl\circ}}$

$\overset{\bullet}{\underset{\bullet\bullet}{Cl\bullet}}$

Arrange the Cl
atoms around C by
sharing electrons
between the two
atoms.

$$Cl-\underset{\underset{Cl}{|}}{\overset{\overset{Cl}{|}}{C}}-Cl$$

(c) $\overset{\times}{\underset{\times}{\times}}\overset{}{C}\times$ $\cdot\overset{\bullet\bullet}{\underset{\circ}{S}}\overset{\bullet}{\circ}$ $\cdot\overset{\circ\circ}{\underset{\circ}{S}}\overset{\circ}{\circ}$

C has 4 valence
electrons and S
has 6.

Placing the S atoms
next to the C atom
does not give 8
electrons around
each atom.

$\overset{\bullet\bullet}{\underset{\bullet}{\circ S}}\overset{\times}{\underset{\times}{\overset{\frown}{C}}}\overset{\circ\circ}{\underset{\circ}{S\circ}}$

Therefore, the single
electrons on each atom
must be shared between
the C and S atoms.

$S = C = S$

There are now two
dashes between
the C and each S
atom - double
bond.

$\overset{\bullet\bullet}{\underset{\bullet\bullet}{\circ S}}\overset{\times\times}{\underset{\circ\circ}{C}}\overset{\times\times}{\underset{\circ\circ}{S}}\overset{\bullet\circ}{\circ}$

(d) $\overset{\times\times}{\underset{\times\times}{\times Cl}}\times$ $\cdot\overset{\circ\circ}{\underset{\circ}{O}}\overset{\circ}{\circ}\cdot\overset{\circ\circ}{\underset{\bullet}{O}}\overset{\circ}{\bullet}\cdot\overset{\circ\circ}{\underset{\bullet}{O}}\overset{\circ}{\bullet}\cdot\overset{\circ\circ}{\underset{\bullet\bullet}{O}}\overset{\circ}{\circ}$ $\overset{\circ\circ}{\underset{\bullet\bullet}{O}}\overset{\circ}{\circ}$

$\overset{\circ\circ}{\underset{\circ\circ}{\circ O}}\overset{\times\times}{\underset{\times\times}{\times Cl}}\overset{\circ\circ}{\underset{\bullet\bullet}{O\otimes}}-$

$\overset{\bullet\bullet}{\underset{\bullet\bullet}{\circ O}}\overset{}{\circ}$

$$\left[\underset{\underset{O}{|}}{\overset{\overset{O}{|}}{O-Cl-O}}\right]^{-}$$

Cl has 7 valence
electrons and O
has 6. There
are 32 bonding
electrons.

$$
\begin{aligned}
1\ Cl &= 7e^- \\
4\ O = 6 \times 4 &= 24e^- \\
1^-\ \text{ionic charge} &= \underline{1e^-} \\
\text{total} &= 3\overline{2e^-}
\end{aligned}
$$

Adding one electron
($\otimes$) due to the charge
on the polyatomic ion
completes the O atom
with 8 electrons.
Place the other 3 O
atoms around the Cl
atom after moving the
single electron to form
a pair on the O atom.
The bond between these O
atoms and the Cl atom is
a coordinate covalent bond.

Draw a ——— for
the coordinate
covalent bond
and disperse the
charge over the
the entire ion.

If your answers are correct, proceed to Problem 6-7. If your
answers are incorrect, review Section 6-7 (Examples 6-2, 6-3,
6-4, 6-5, 6-6, 6-7, 6-8, and 6-9 and Study Exercise 6-6) in
your text.

Proceed to Problem 6-6.

70

PROBLEM 6-6

Write Lewis structures and structural formulas for the following molecules or polyatomic ions:

(a) CH_4 (a) _____

(b) C_2H_4 (b) _____

(c) N_2 (c) _____

(d) $PO_4{}^{3-}$ (d) _____

ANSWERS AND SOLUTIONS TO PROBLEM 6-6

Electron-dot formulas of elements	Electron-dot formulas of molecules or ions	Structural formulas

(a) $\overset{\times}{\underset{\times}{\times}}\!C\!\times$ •H•H•H°H

C has 4 valence electrons and H has 1.

```
        H
        ×°
  H ⦁ C ⦁ H
        ×°
        H
```

C obeys the rule of 8 and H obeys the rule of 2. Placing the H atoms around the C atom obeys these rules.

```
        H
        |
  H —— C —— H
        |
        H
```

The bond angle is 109.5°, but you need not be concerned with this now.

(b) $\times\!\overset{\times}{\underset{\times}{C}}\!\times$ $\times\!\overset{\times}{\underset{\times}{C}}\!\times$•H   •H

•H   •H

Placing the C atoms together and adding H atoms to the two C atoms does not give 8 electrons around each C atoms.

```
       ×
  H ⦁ C ⦇ × C ⦁ H
      ×°   ×°
      H     H
```

Therefore, the single electrons around each C atom must be shared.

```
      ×      ××      ×
  H ⦁ C × × C ⦁ H
      ×⦁    ×⦁
      H      H
```

Using two dashes between the C atoms forms a double bond. The bond angle is 120°, but this will be covered in Chapter 18.

71

(c) $\overset{\times}{\underset{\times}{\times}} \text{N}^{\times}$ $\bullet \overset{\bullet}{\underset{\bullet}{\text{N}}} \overset{\bullet}{\bullet}$

N has 5 valence
electrons.

Placing the N atoms
together does not give
8 electrons around
each N atom.

$\overset{\times}{\underset{\times}{\times}} \text{N}^{\times} \overset{\bullet}{\underset{\bullet}{\bullet}} \text{N} \overset{\bullet}{\bullet}$

Therefore, the single N
electrons around each N
atom must be shared
between N atoms.

$\overset{\times}{\underset{\times}{\times}} \text{N} \overset{\times}{\underset{\times}{\times}} \overset{\bullet}{\underset{\bullet}{\bullet}} \text{N} \overset{\bullet}{\underset{\bullet}{\bullet}}$

N$\equiv$N

Three dashes be-
tween the N atoms
shows a triple
bond.

(d) $\overset{\times\times}{\underset{\times}{\times}} \text{P}^{\times}$ $\bullet \overset{\bullet\bullet}{\text{O}} \overset{\bullet}{\underset{\bullet}{\bullet}}$ $\bullet \overset{\bullet\bullet}{\text{O}} \overset{\bullet}{\underset{\bullet}{\bullet}}$ $\bullet \overset{\bullet\bullet}{\text{O}} \overset{\bullet}{\underset{\bullet}{\bullet}}$ $\bullet \overset{\bullet\bullet}{\text{O}} \overset{\bullet}{\underset{\bullet}{\bullet}}$ $\overset{\bullet\bullet}{\bullet\text{O}\bullet}$

P has 5 valence
electrons and O
has 6. There
are 32 bonding
electrons.

  1 P = 5e⁻
4 O = 6 x 4 = 24e⁻
3⁻ ionic   =   3e⁻
  charge
    total = 32e⁻

Placing the O atoms
around the P atom and
adding 3 electrons
(⊗) due to the charge
on the polyatomic ion
gives 8 electrons
around each atom. Also,
placing an O atom next
to the P atom where both
electrons are contributed
by the P atom. This bond
is a coordinate covalent
bond.

$-\overset{}{\underset{}{\bigotimes}} \text{O} \overset{\bullet\bullet}{\underset{\bullet\bullet}{\text{X}}} \text{P} \overset{\times\times}{\underset{\times\bullet}{\times}} \text{O} \overset{}{\underset{}{\bigotimes}} -$

$\overset{\bullet\bullet}{\bullet \text{O} \bullet} \overset{}{\bigotimes} -$

Draw a ———for the
single coordinate
covalent bond and
disperse the charge
over the entire
ion.

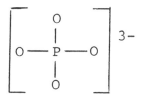

If your answers are correct, proceed to Problem 6-7. If your
answers are incorrect, re-read Section 6-7 in your text. Re-
work Problem 6-5 and 6-6.

Proceed to Problem 6-7.

PROBLEM 6-7

Determine the shape (linear, trigonal planar, tetrahedral,
bent, or pyramidal) and the approximate bond angle of the
following molecules: (You may use the periodic table.)

(a) $CHCl_3$

(a) _____

(b) H_2CO

(b) _____

ANSWERS AND SOLUTIONS TO PROBLEM 6-7

(a) tetrahedral, 109.5°

Using the periodic table, C has 4 valence electrons, H has
1, and Cl has 7 each, so the Lewis structure (Guideline 1)
for $CHCl_3$ is:

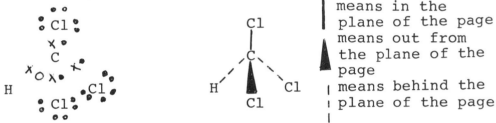

The total number of electron pairs or groups (Guideline 2)
is 4. Obeying Coulomb's law (Guideline 3), the structures
are as follows:

	means in the plane of the page
	means out from the plane of the page
	means behind the plane of the page

With 4 bonding pairs and 0 unshared pairs (Table 6-6), the
shape is tetrahedral with a bond angle of 109.5°.

(b) trigonal planar, 120°

Using the periodic table, C has 4 valence electrons, H has
1 each, and O has 6, so the Lewis structure (Guideline 1)
for H_2CO is:

(In order to obey the rule of 8 for both C and O, a double
bond must be formed between the two atoms.)

The total number of groups (Guideline 2) is 3 with 2 pairs
of elctrons in the C-H single bond and a group of 4
electrons in the C═O double bond. Obeying Coulomb's law
(Guideline 3), the structures are as follows:

73

With 3 bonding groups and 0 unshared pairs around carbon (Table 6-6), the shape is trigonal planar with a bond angle of 120°.

If your answers are correct, proceed to Problem 6-8. If your answers are incorrect, review Section 6-8 (Example 6-10 and Study Exercise 6-7) in your text.

Proceed to Problem 6-8.

PROBLEM 6-8

Write the correct formulas for the compound formed by the combination of the following ions:

(a) calcium (Ca^{2+}) and sulfide (S^{2-})   (a) _____

(b) calcium (Ca^{2+}) and chloride (Cl^-)   (b) _____

(c) calcium (Ca^{2+}) and nitride (N^{3-})   (c) _____

(d) calcium (Ca^{2+}) and phosphate ($PO_4{}^{3-}$)   (d) _____

ANSWERS AND SOLUTION TO PROBLEM 6-8

(a) CaS $(Ca^{2+}) (S^{2-})$, CaS

$$2^+ + 2^- = 0$$

(b) $CaCl_2$ $(Ca^{2+}) (Cl^-)_2$, $CaCl_2$

$$2^+ + 2(1^-) = 0$$

(c) Ca_3N_2 $(Ca^{2+})_3 (N^{3-})_2$, Ca_3N_2

$$3(2^+) + 2(3^-) = 6^+ + 6^- = 0$$

The least common multiple is 6 - hence 3(2) and 2(3).

(d) $Ca_3(PO_4)_2$ $(Ca^{2+})_3 (PO_4{}^{3-})_2$, $Ca_3(PO_4)_2$

$$3(2^+) + 2(3^-) = 6^+ + 6^- = 0$$

Because there are two phosphate ions, you must use the ( ). The ( ) means that there are 2 atoms of phosphorus, 8 atoms of oxygen, and 3 atoms of calcium in one formula unit of calcium phosphate.

If your answers are correct, proceed to Problem 6-9. If your answers are incorrect, review Section 6-9 (Study Exercise 6-8) in your text.

Proceed to Problem 6-9.

PROBLEM 6-9

Using the periodic table, indicate a maximum positive oxidation number for each of the following elements. For those elements that are nonmetals, give both the maximum positive oxidation number and the negative oxidation number. (If you do not know the symbol for the element, look it up inside the front cover of your text.)

(a) strontium   (a) _____

(b) gallium   (b) _____

(c) phosphorus   (c) _____

(d) tellurium   (d) _____

ANSWERS AND SOLUTIONS TO PROBLEM 6-9

(a) 2^+ Strontium (Sr) is in group IIA (2) and has a 2^+ maximum positive oxidation number.

(b) 3^+ Gallium (Ga) is in group IIIA (13) and has a 3^+ maximum positive oxidation number.

(c) 5^+ and 3^- Phosphorus (P) is in group VA (15) and has a 5^+ maximum postive oxidation number and a 3^- negative oxidation number (V - 8 = - 3).

(d) 6^+ and 2^- Tellurium (Te) is in group VIA (16) and has a 6^+ maximum positive oxidation number and a 2^- negative oxidation number (VI - 8 = -2).

If your answers are correct, proceed to Problem 6-10. If your answers are incorrect, review Section 6-10 - Oxidation Numbers - (Study Exercise 6-9) in your text.

Proceed to Problem 6-10.

PROBLEM 6-10

Using the periodic table to determine the oxidation numbers, predict the formula of the binary compound formed from the following combinations of elements. (If you do not know the symbol for the element, look it up inside the front cover of your text.)

(a) magnesium and chlorine   (a) _____

(b) magnesium and nitrogen   (b) _____

(c) potassium and sulfur   (c) _____

(d) aluminum and sulfur   (d) _____

ANSWERS AND SOLUTIONS TO PROBLEM 6-10

(a) $MgCl_2$ Magnesium (Mg) is in group IIA (2) and has a 2^+ maximum positive oxidation number; chlorine (Cl) is in group VIIA (17) and has a 1^- negative oxidation number. The correct formula is $(Mg^{2+})(Cl^-)_2$, $MgCl_2$.

(b) Mg_3N_2 Mg has a 2^+ maximum positive oxidation number. Nitrogen (N) is in group VA (15) and has a 3^- negative oxidation number (V - 8 = -3). The correct formula is $(Mg^{2+})_3(N^{3-})_2$, Mg_3N_2.

(c) K_2S Potassium (K) is in group IA (1) and has a 1^+ maximum positive oxidation number; sulfur is in group VIA (16) and has a 2^- negative oxidation number. The correct formula is $(K^+)_2(S^{2-})$, K_2S.

(d) Al_2S_3 Aluminum (Al) is in group IIIA (13) and has a 3^+ maximum positive oxidation number; S has a 2^- negative oxidation number. The correct formula is $(Al^{3+})_2(S^{2-})_3$, Al_2S_3.

If your answers are correct, proceed to Problem 6-11. If your answers are incorrect, review Section 6-10 - Oxidation Numbers - (Study Exercise 6-10) in your text.

Proceed to 6-11.

PROBLEM 6-11

Estimate the missing value for the following:

(a) Element   Boiling Point ($^{\circ}$C, 1 atm)

    Li   1372
    Na   880
    K   776
    Rb   ?   (a) _____

(b) Element   Radius (pm)

    N   75.0
    P   11$\overline{0}$
    As   122
    Sb   ?   (b) _____

ANSWERS AND SOLUTIONS TO PROBLEM 6-11

(a) 672°C   The boiling point decreases as we move down the
              group, so we would expect the boiling point of
              Rb to be less than that of K. You can estimate
              this value by taking the difference between the
              last two values given and subtracting it from
              the value for K.

$$776^{\circ}C - (880 - 776)^{\circ}C = 672^{\circ}C$$
The actual value is 679°C.

(b) 134 pm   The radius increases as we move down the group,
              due to the addition of a new principal energy
              level; hence, we would expect the radius of Sb
              to increase. Estimate the value by taking the
              difference between the last two values given and
              adding it to the value for As.

$$122 \text{ pm} + (122 - 110) \text{ pm} = 134 \text{ pm}$$
The actual value is 143 pm.

If your answers are correct, proceed to Problem 6-12. If your
answers are incorrect, review Section 6-10 - Properties -
(Study Exercise 6-11) in your text.

Proceed to Problem 6-12.

PROBLEM 6-12

The following are some examples of compounds and their formulas:

calcium chloride, $CaCl_2$

calcium phosphate, $Ca_3(PO_4)_2$

sodium bromide, NaBr

Using the periodic table, write the formulas for the following compounds. (<u>Hint</u>: Note the endings of each name for the compound.)

(a) potassium bromide   (a) _____

(b) magnesium phosphate   (b) _____

(c) magnesium iodide   (c) _____

(d) calcium arsenate   (d) _____

ANSWERS AND SOLUTIONS TO PROBLEM 6-12

(a) KBr   Potassium bromide is comparable to sodium bromide, because potassium is in the same group as sodium. Potassium bromide is not comparable to calcium chloride because calcium and potassium are not in the same group.

(b) $Mg_3(PO_4)_2$ Magnesium phosphate is comparable to calcium phosphate, because magnesium is in the same group as calcium. Magnesium phosphate is not comparable to either calcium chloride or sodium bromide, as both of these compounds have -<u>ide</u> endings and magnesium phosphate has an -<u>ate</u> ending.

(c) MgI_2 Magnesium iodide is comparable to calcium chloride, because magnesium is in the same group as calcium and iodine is in the same group as chlorine.

(d) $Ca_3(AsO_4)_2$ Calcium arsenate is comparable to calcium phosphate, because arsenic is in the same group as phosphorus. Notice that the ending in calcium arsenate is -<u>ate</u> and so calcium arsenate is not comparable to either calcium chloride or sodium bromide which both have -<u>ide</u> endings.

If your answers are correct, proceed to Problem 6-13. If your answers are incorrect, review Section 6-10 - Formulas - (Study Exercise 6-12) in your text.

Proceed to Problem 6-13.

PROBLEM 6-13

Using the periodic table, classify the following compounds as essentially ionic or covalent.

(a) NaCl   (a) _____

(b) SO_2 (b) _____

(c) CuF_2 (c) _____

(d) PCl_3 (d) _____

(e) $Ca_3(PO_4)_2$ (e) _____

(f) $Ba(NO_3)_2$ (f) _____

ANSWERS AND SOLUTIONS TO PROBLEM 6-13

(a) ionic   NaCl is an ionic compound, because Na is in group IA (1) and Cl is in group VIIA (17).

(b) covalent SO_2 is a covalent compound, because S and O are nonmetals.

(c) ionic CuF_2 is an ionic compound, because Cu is a metal and any metal with F is ionic.

(d) covalent PCl_3 is a covalent compound, because P and Cl nonmetals.

(e) ionic $Ca_3(PO_4)_2$ is an ionic compound, because PO_4^{3-} is a polyatomic ion.

(f) ionic $Ba(NO_3)_2$ is an ionic compound, because NO_3^- is a polyatomic ion.

If your answers are correct, you have completed this chapter. If your answers are incorrect, review Section 6-10 - Types of Bonding - Ionic or Covalent (Study Exercise 6-13) in your text.

These problems conclude the chapter on the structure of compounds.

Now take the sample quiz to see if you have mastered the material in this chapter.

SAMPLE QUIZ

Quiz #6 You may use the periodic table and the list of symbols
        of the elements.

1. Diagram the ionic structure for each of the following ions,
    indicating the number of protons and neutrons in the
    nucleus and arranging the electrons in principal energy
    levels.

    (a) $^{39}_{19}K^+$ _____

    (b) $^{19}_{9}F^-$ _____

2. Write the correct formulas for the compound formed by the
    combination of the following ions:

    (a) barium (Ba^{2+}) and chloride (Cl^-) _____

    (b) calcium (Ca^{2+}) and nitrate (NO_3^-) _____

    (c) iron(III) (Fe^{3+}) and oxalate $(C_2O_4^{2-})$ _____

    (d) tin(IV) (Sn^{4+}) and chromate (CrO_4^{2-}) _____

3. Write Lewis structures and structural formulas for the
    following:

    (a) HF

    (b) N_2

4. Estimate the missing value:

Element	Boiling Point ($^{\circ}$C, 1 atm)
Fe	2750
Ru	3900
Os	?

5. Classify the following compounds as essentially ionic or
    covalent:

    (a) CO_2 _____   (b) CdF_2 _____

    (c) NaI _____   (d) $Mg(NO_3)_2$ _____

6. Calcuate the oxidation number of the element indicated in each of the following compounds or ions:

(a) Si in SiO_2

(b) Si in $SiO_3{}^{2-}$

(a) _____

(b) _____

Solutions and Answers for Quiz #6

1. (a) (19p 20n) $2e^-$ $8e^-$ $8e^-$ 1^+

   (b) (9p 10n) $2e^-$ $8e^-$ 1^-

2. (a) $BaCl_2$; (b) $Ca(NO_3)_2$; (c) $Fe_2(C_2O_4)_3$; (d) $Sn(CrO_4)_2$

3. (a) H ⋅ F̈   H—F

   (b) N⋮⋮N   N≡N

4. $39\overline{0}0°C + (39\overline{0}0 - 2750)°C = 5050°C$

   The actual value is $5027°C$.

5. (a) covalent; (b) ionic; (c) ionic; (d) ionic

6. (a) ox. no. Si + 2(-2) = 0   (b) ox. no. Si + 3(-2) = -2
      ox. no. Si - 4 = 0   ox. no. Si - 6 = -2
      ox. no. Si = + 4   ox. no. Si = 6 - 2
                                        ox. no. Si = + 4

Review Exam #2 (Chapters 4 to 6) [Answers are in ( ) to the right of each question.]

1. The nucleus of an atom is:

   A.   Positively charged and has a high density   (A)

   B.   Negatively charged and has a high density

   C.   Positively charged and has a low density

   D.   Negatively charged and has a low density

   E.   None of these

2. Two naturally occurring isotopes of the element iridium (Ir) are $^{191}_{77}$Ir and $^{193}_{77}$Ir. How many neutrons are present in the nucleus of each isotope?

   A. 191 and 193, respectively   (B)

   B. 114 and 116, respectively

   C. 77 in both

   D. 192 in both

   E. 268 and 270, respectively

3. Given an isotope of the element silicon, $^{30}_{14}$Si, the number of electrons in the principal energy levels, 1, 2, and 3 are:

   A. 2, 8, 8, respectively, plus 8 in 4 and 4 in 5   (C)

   B. 2, 8, 6, respectively

   C. 2, 8, 4, respectively

   D. 2, 8, 3, respectively

   E. None of these

4. The correct representation of the electronic structure in sublevels for $^{34}_{16}$S is:

   A. $1s^2, 2s^2\ 2p^6, 3s^2\ 3p^4$ (A)

   B. $1s^2, 2s^2\ 2p^6, 3s^2\ 3p^5$

   C. $1s^2, 2s^2\ 2p^6, 3s^2\ 3p^5$

   D. $1s^2, 2s^2\ 2p^6, 3s^2\ 3p^6, 4s^2\ 4p^4$

   E. None of these

5. The electron-dot formula for $^{27}_{13}$Al is:

A. Al $\overset{\bullet}{}$ (C)

B. $\overset{\bullet}{\text{Al}}$

C. $\overset{\circ}{\underset{\bullet}{\text{Al}}}$ $\bullet$

D. $\overset{\bullet\,\bullet}{\underset{\bullet\,\bullet}{\,\bullet\,\text{Al}\,\circ}}$

E. Al $\overset{\bullet}{\underset{\bullet}{}}\bullet$

For questions 6 to 16 refer to the periodic table.

6. The number of valence electrons in $^{37}_{17}$Cl is:

A. 3   (E)

B. 4

C. 5

D. 6

E. 7

7. Which one of the following is the ionic charge of $^{80}_{34}$Se when it exists as a negative ion?

A. 3^- (B)

B. 2^-

C. 1^-

D. 4^-

E. 5^-

8. Which one of the following is <u>not</u> a correct statement about $^{40}_{20}$Ca?

   A.   It is metallic in character.   (D)

   B.   It has two valence electrons.

   C.   It has two 4s electrons.

   D.   It has two 3d electrons.

   E.   It has 20 electrons outside the nucleus.

9. An element whose electronic structure is represented as

   $1\underline{s}^2$, $2\underline{s}^2$ $2\underline{p}^6$,$3\underline{s}^2$ $3\underline{p}^6$ $3\underline{d}^2$, $4\underline{s}^2$ belongs to the class of
   elements known as:

   A.   the alkali metals   (D)

   B.   the alkaline earth metals

   C.   the halogens

   D.   the transition elements

   E.   the A group elements

10. In which series of elements would the size of the atom
    increase?

   A.   N, O, F   (D)

   B.   Kr, Ar, Ne

   C.   Sc, Ti, V

   D.   Li, Na, K

   E.   Na, Mg, Al

11. The compound formed by $^{23}_{11}$Na and $^{81}_{35}$Br would probably

   A.   be an ionic compound.   (A)

   B.   have the formula $NaBr_2$.

   C.   be a covalent compound.

   D.   be a coordinate covalent compound.

   E.   have a low melting point.

84

12. An <u>ion</u> of $^{39}_{19}$K will have

  A. a negative charge.   (C)

  B. two (2) neutrons in principal energy level 1.

  C. a total of 18 electrons.

  D. no neutrons.

  E. one (1) electron in principal energy level 4.

13. Which one of the following elements is the most metallic?

  A. Al   (E)

  B. Si

  C. P

  D. S

  E. Ga

14. Estimate the missing value:

Element	Density (g/L, gas, 0°C and 1 atm)
Ne	0.900
Ar	1.78
Kr	3.73
Xe	?

  A. 2.76 g/L   (E)

  B. 2.85 g/L

  C. 4.61 g/L

  D. 1.46 g/L

  E. 5.68 g/L

15. Estimate the missing value:

Compound	Density of solid (g/mL)
SF_6	2.51
SeF_6	?
TeF_6	3.76

A.   2.64 g/mL   (B)

B.   3.14 g/mL

C.   5.29 g/mL

D.   2.14 g/mL

E.   1.26 g/mL

16. Classify the following compounds as essentially ionic or covalent:

$$Co(NO_3)_2, \ CoO, \ CO_2$$

A.   All are ionic   (C)

B.   All are covalent

C. $Co(NO_3)_2$ and CoO are ionic; CO_2 is covalent

D. $Co(NO_3)_2$ and CoO are covalent; CO_2 is ionic

E. CoO and CO_2 are covalent; $Co(NO_3)_2$ is ionic

17. Write the electronic configuration in sublevels for $^{31}_{15}P^{3-}$.

A. $1s^2, 2s^2 2p^6, 3s^2 3p^3$ (C)

B. $1s^2, 2s^2 2p^6 3s^2 3p^5$

C. $1s^2, 2s^2 2p^6, 3s^2 3p^6$

D. $1s^2, 2s^2 2p^6, 3s^2$

E.   None of these

18. Using Lewis structures, show which one of the following is the correct electronic structure for water.

$$(^1_1H \text{ and } ^{16}_8O)$$

A. 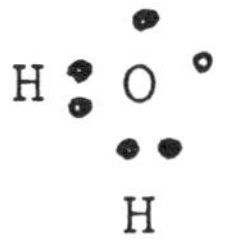 H ⦂ O ˚

    H

B. 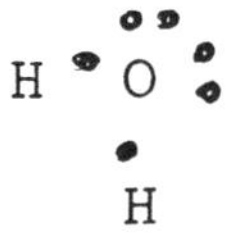 H ⦂ O ⦂

    H

C. 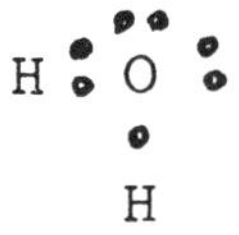 H ⦂ O ⦂

    H

D. 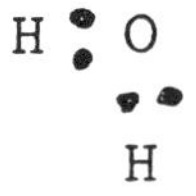 H ⦂ O

    H

E. 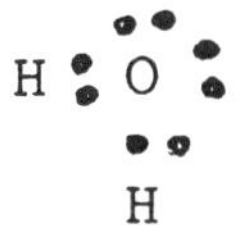 H ⦂ O ⦂

    H
                                                    (E)

19. Using Lewis structures, show which one of the following is the correct structure for chlorate ion (ClO_3^-).

$$\left(^{35}_{17}Cl, \quad ^{16}_{8}O\right)$$

A.  :Cl: O: (D)

B.  :Cl: O: −

C. 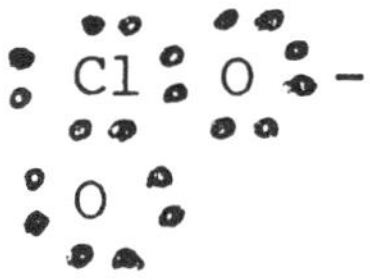 : Cl: O: −
    : O:

D. 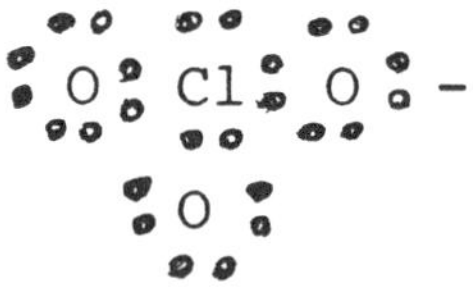 :O: Cl: O: −
      :O:

E. 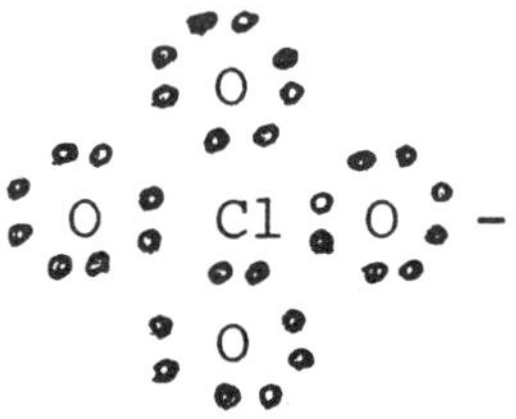   :O:
    :O: Cl: O: −
      :O:

20. Using a dash (——) to represent a pair of electrons, show which one of the following is the correct structure for N_2 $\left(^{14}_{7}N\right)$.

A.   N——N   (C)

B.   N══N

C.   N≡══N

D.   N≣══N

E.   None of these

21. Determine the formula of the compound composed of strontium (Sr^{2+}) and phosphate (PO_4^{3-}) ions.

    A. $SrPO_4$ (D)

    B. Sr_3PO_4

    C. $Sr_2(PO_4)_2$

    D. $Sr_3(PO_4)_2$

    E. $Sr_3P_2O_7$

22. Determine the formula of the compound composed of tin(IV) (Sn^{4+}) and sulfate (SO_4^{2-}) ions.

    A. $Sn_2(SO_4)_4$ (B)

    B. $Sn(SO_4)_2$

    C. $SnSO_4$

    D. SnS_2O_7

    E. Sn_2SO_4

23. Determine the formula of the compound composed of ferric (Fe^{3+}) and chloride (Cl^-) ions.

    A. $FeCl$ (D)

    B. Fe_2Cl

    C. Fe_3Cl

    D. $FeCl_3$

    E. $FeCl_2$

24. Calculate the maximum number of electrons that may occupy principal energy level 7.

    A.   72   (B)

    B.   98

    C.   50

    D.   68

    E.   49

25. Calculate the atomic mass to four significant digits for the hypothetical element X, given the following data:

Isotope	Exact atomic mass in amu	% Abundance in Nature
^{15}X	15.00	30.00
^{17}X	17.10	70.00

A. 16.05 amu   (E)

B. 15.63 amu

C. 16.00 amu

D. 16.50 amu

E. 16.47 amu

CHAPTER 7

CHEMICAL NOMENCLATURE OF INORGANIC COMPOUNDS

In Chapter 7 we will consider formula writing and naming inorganic compounds. To do this you must know Tables 6-1, 6-2, and 6-5. The periodic table is of great help to you in determining the ionic charge of the cations and anions in Table 6-1 and 6-2. See Chapter 6 in this book (Selected Topic number 2).

Chemical nomenclature is divided into systematic chemical names and common names. The systematic chemical names are used most often, but some common names, such as water - H_2O, still persist. We will consider primarily systematic chemical names in this chapter.

SELECTED TOPICS

1. Binary compounds contain two different elements. The more positive portion - that is, the metal, the positive polyatomic ion, the hydrogen ion, or the less electronegative nonmetal is named and written first. The more negative portion - that is, the more electronegative nonmetal or negative polyatomic ion is named and written last. See Section 7-1 in your text.

2. For all binary compounds, the ending of the second element is -ide. When both elements are nonmetals (see periodic table), the number of atoms of each element is indicated in the name with Greek prefixes (see Table 7-1 in your text), except in the case of mono (one), which is used for only the second nonmetal. See Section 7-2 in your text.

3. In naming binary compounds containing a metal and a nonmetal Greek prefixes are not used. For those binary compounds containing a metal with a fixed charge, the charge is not stated in the name. For those binary compounds containing a metal with a variable charge, the charge is indicated in the name by one of two methods: the Stock system or the -ous or -ic suffix system. In the Stock system, the charge of the metal is indicated by a Roman numeral in parentheses immediately following the name of the metal. In the -ous or -ic suffix method, the Latin stem for the metal is used with an -ous or -ic suffix. The -ous represents the lower charge and the -ic the higher charge. To write the formulas of binary compounds, you must know or be able to determine the ionic charges of the cations or anions given in Table 6-1 and 6-2. See Section 7-3 in your text.

4. <u>Ternary</u> <u>compounds</u> are composed of three or more types of atoms. In naming and writing formulas of ternary and higher compounds, the same procedure is used as for binary compounds, except that the name or formula of the polyatomic ion is used. You must memorize the names and formulas of the polyatomic ions given in Table 6-5. Many of the polyatomic ions in Table 6-5 have endings of -<u>ate</u> or -<u>ite</u>. The -<u>ate</u> ion has one more oxygen atom than the -<u>ite</u>

   ion. For example, the formula for nitrite is NO_2^-, and

   the formula for nitrate is NO_3^-. Therefore, if you know

   the formula of one of these ions, you know the formula of the other one. Three polyatomic ions in the table do not

   have -<u>ate</u> or -<u>ite</u> endings. They are: ammonium ion - NH_4^+,

   hydroxide - OH^-, and cyanide - CN^-. See Section 7-4 in your text.

5. Table 6-4 lists four different polyatomic ions containing

   chlorine: perchlorate - ClO_4^-, chlorate - ClO_3^-, chlorite

   - ClO_2^-, and hypochlorite - ClO^-. The prefix <u>hypo</u>- means

   "under", so hypochlorite has one atom "under" the number of oxygen atoms of chlorite. The prefix <u>per</u>- means "over", so perchlorate has one atom "over" the number of oxygen atoms of chlorate. These prefixes can be used for other oxy-halogen ions, such as those derived from bromine and iodine, but not fluorine. Fluorine does not form polyatomic ions with oxygen because both elements are too highly electronegative and the ions are not stable. See Section 7-5 in your text.

6. Acid and bases are special types of compounds. <u>Acids</u> are

   hydrogen compounds that yield hydrogen ions (H^+) in aqueous (water) solution. <u>Bases</u> are compounds that contain

   a metal ion and a hydroxide ion (OH^-). A <u>salt</u> is an ionic compound made up of a positively charged ion (cation) and a negatively charged ion (anion).

   Given the formula of a compound, you should be able to classify it as an acid, base, or salt. The following are generalized formulas for acids, bases, or salts.

HX = ACID   [H is a hydrogen ion and X is an anion
             (nonmetal ion or negative polyatomic ion)
             in aqueous (water) solution.]

MOH = BASE   (M is a metal ion and OH is a hydroxide
             ion.)

MX = SALT   [M is a cation (metal ion or positive
             polyatomic ion) and X is an anion
             (nonmetal ion or negative polyatomic
             ion.)]

See Section 7-6 in your text.

7. Table 7-4 in your text gives some common names and their
    formulas. You should memorize the formulas of these com-
    pounds and their common names. See Section 7-7 in your
    text.

PROBLEM 7-1

Write the correct name for each of the following compounds:

(a) SO_2 (a) _____

(b) CO_2 (b) _____

(c) ClO_2 (c) _____

(d) N_2O_5 (d) _____

(e) Cl_2O_7 (e) _____

ANSWERS TO PROBLEM 7-1

(a) sulfur dioxide

(b) carbon dioxide

(c) chlorine dioxide

(d) dinitrogen pentoxide

(e) dichlorine heptoxide

If your answers are correct, proceed to Problem 7-2. If your
answers are incorrect, review Section 7-2 (Study Exercise 7-1)
in your text.

Proceed to Problem 7-2.

PROBLEM 7-2

Write the correct formula for each of the following compounds:

(a) phosphorus trichloride   (a) _____

(b) phosphorus pentachloride   (b) _____

(c) carbon monoxide   (c) _____

(d) sulfur trioxide   (d) _____

(e) dinitrogen trioxide   (e) _____

ANSWERS TO PROBLEM 7-2

(a) PCl_3

(b) PCl_5

(c) CO

(d) SO_3

(e) N_2O_3

If your answers are correct, proceed to Problem 7-3. If your answers are incorrect, review Section 7-2 (Study Exercise 7-2) in your text.

Proceed to Problem 7-3.

PROBLEM 7-3

Write the correct name for each of the following compounds:

(a) $AgCl$ (a) _____

(b) Al_2O_3 (b) _____

(c) Mg_3N_2 (c) _____

(d) $HgCl_2$ (d) _____

(e) SnF_2 (e) _____

ANSWERS TO PROBLEM 7-3

(a) silver chloride

(b) aluminum oxide (No Greek prefixes)

(c) magnesium nitride

(d) mercury(II) chloride or mercuric chloride (chloride is 1^-, Hg is 2^+)

(e) tin(II) fluoride or stannous fluoride (fluoride is 1^-, Sn is 2^+)

If your answers are correct, proceed to Problem 7-4. If your answers are incorrect, review Section 7-3 (Study Exercises 7-3 and 7-5) in your text.

Proceed to Problem 7-4.

PROBLEM 7-4

Write the correct formula for each of the following compounds:

(a) potassium bromide   (a) _____

(b) calcium fluoride   (b) _____

(c) copper(II) chloride   (c) _____

(d) iron(III) oxide   (d) _____

(e) tin(II) iodide   (e) _____

ANSWERS TO PROBLEM 7-4

(a) KBr (K is 1^+; Br is 1^-; see Tables 6-1 and 6-2)

(b) CaF_2 (Ca is 2^+; F is 1^-

(c) $CuCl_2$ (Cu is 2^+; Cl is 1^-)

(d) Fe_2O_3 (Fe is 3^+; O is 2^-)

(e) SnI_2 (Sn is 2^+; I is 1^-)

If your answers are correct, proceed to Problem 7-5. If your answers are incorrect, review Section 7-3 (Study Exercises 7-4 and 7-6) in your text.

Proceed to Problem 7-5.

PROBLEM 7-5

Write the correct name for each of the following compounds:

(a) $Sr(NO_3)_2$ (a) _____

(b) AgCN   (b) _____

(c) $Ca(HSO_4)_2$ (c) _____

95

(d) $FeCO_3$ (d) _____

(e) $Sn(NO_3)_4$ (e) _____

ANSWERS TO PROBLEM 7-5

(a)   strontium nitrate

(b)   silver cyanide

(c)   calcium hydrogen sulfate or calcium bisulfate

(d)   iron(II) carbonate or ferrous carbonate (carbonate is

      2^-, Fe is 2^+)

(e)   tin(IV) nitrate or stannic nitrate (nitrate is 1^-, Sn

      is 4^+)

If your answers are correct, proceed to Problem 7-6. If your
answers are incorrect, review Section 7-4 (Study Exercise 7-7)
in your text.

Proceed to Problem 7-6.

PROBLEM 7-6

Write the correct formula for each of the following compounds:

(a)   sodium chromate   (a) _____

(b)   calcium permanganate   (b) _____

(c)   potassium oxalate   (c) _____

(d)   mercury(II) nitrate   (d) _____

(e)   ammonium sulfate   (e) _____

ANSWERS TO PROBLEM 7-6

(a) Na_2CrO_4 (Na is 1^+; CrO_4 is 2^-; see Tables 6-1 and 6-4)

(b) $Ca(MnO_4)_2$ (Ca is 2^+; MnO_4 is 1^-)

(c) $K_2C_2O_4$ (K is 1^+; C_2O_4 is 2^-)

(d) $Hg(NO_3)_2$ (Hg is 2^+; NO_3 is 1^-)

(e) $(NH_4)_2SO_4$ (NH_4 is 1^+; SO_4 is 2^-)

If your answers are correct, proceed to Problem 7-7. If your answers are incorrect, review Section 7-4 (Study Exercise 7-8) in your text.

Proceed to Problem 7-7.

PROBLEM 7-7

Write the correct name for each of the following compounds:

(a) KClO   (a) _____

(b) $Cd(ClO_3)_2$ (b) _____

(c) $Cu(ClO_4)_2$ (c) _____

(d) $Pb(IO_3)_2$ (d) _____

(e) $Hg(ClO_3)_2$ (e) _____

ANSWERS TO PROBLEM 7-7

(a) potassium hypochlorite

(b) cadmium chlorate

(c) copper(II) perchlorate or cupric perchlorate (perchlorate is 1^-, Cu is 2^+)

(d) lead(II) iodate or plumbous iodate (iodate is 1^-, Pb is 2^+)

(e) mercury(II) chlorate or mercuric chlorate (chlorate is 1^-, Hg is 2^+)

If your answers are correct, proceed to Problem 7-8. If your answers are incorrect, review Section 7-5 (Study Exercise 7-9) in your text.

Proceed to Problem 7-8.

PROBLEM 7-8

Write the correct formula for each of the following compounds:

(a) calcium perchlorate   (a) _____

(b) potassium iodite   (b) _____

(c) barium hypochlorite   (c) _____

(d) lead(II) bromate   (d) _____

97

(e) iron(III) iodate   (e) _____

ANSWERS TO PROBLEM 7-8

(a) $Ca(ClO_4)_2$ (Ca is 2^+; ClO_4 is 1^-; see Tables 6-1 and 6-4)

(b) KIO_2 (K is 1^+; IO_2 is 1^-)

(c) $Ba(ClO)_2$ (Ba is 2^+; ClO is 1^-)

(d) $Pb(BrO_3)_2$ (Pb is 2^+; BrO_3 is 1^-)

(e) $Fe(IO_3)_3$ (Fe is 3^+; IO_3 is 1^-)

If your answers are correct, proceed to Problem 7-9. If your answers are incorrect, review Section 7-5 (Study Exercise 7-10) in your text.

Proceed to Problem 7-9.

PROBLEM 7-9

Write the correct name for the following compounds:

(a) $Al(OH)_3$ (a) _____

(b) $HClO_4$ in aqueous (b) _____
     solution

(c) H_2CrO_4 in aqueous (c) _____
     solution

(d) $Fe(OH)_2$ (d) _____

(e) $Ba(OH)_2$ (e) _____

ANSWERS TO PROBLEM 7-9

(a) aluminum hydroxide

(b) perchloric acid

(c) chromic acid

(d) iron(II) hydroxide or ferrous hydroxide (hydroxide is 1^-, Fe is 2^+)

(e) barium hydroxide (hydroxide is 1^-, barium is 2^+)

If your answers are correct, proceed to Problem 7-10. If your answers are incorrect, review Section 7-6 (Study Exercises 7-11 and 7-13) in your text.

Proceed to Problem 7-10.

PROBLEM 7-10

Write the correct formulas for the following compounds:

(a) nitric acid   (a) _____

(b) iron(III) hydroxide   (b) _____

(c) phosphoric acid   (c) _____

(d) lead(II) hydroxide   (d) _____

(e) sulfuric acid   (e) _____

ANSWERS TO PROBLEM 7-10

(a) HNO_3 (H is 1^+; NO_3 is 1^-; see Tables 6-1 and 6-4)

(b) $Fe(OH)_3$ (Fe is 3^+; OH is 1^-)

(c) H_3PO_4 (H is 1^+; PO_4 is 3^-)

(d) $Pb(OH)_2$ (Pb is 2^+; OH is 1^-)

(e) H_2SO_4 (H is 1^+; SO_4 is 2^-)

If your answers are correct, proceed to Problem 7-11.   If your answers are incorrect, review Section 7-6 (Study Exercises 7-12 and 7-14) in your text.

Proceed to Problem 7-11.

PROBLEM 7-11

Classify each of the following compounds as (1) an acid, (2) a base, or (3) a salt. Assume that all soluble compounds are in aqueous solution.

(a) $Cu(CN)_2$ (a) _____

(b) $Pt(OH)_2$ (b) _____

(c) H_2CrO_4 (c) _____

(d) Li_2SO_4 (d) _____

(e) $Mg(OH)_2$ (e) _____

ANSWERS TO PROBLEM 7-11

(a) (3), salt

(b) (2), base

(c) (1), acid

(d) (3), salt

(e) (2), base

If your answers are correct, proceed to Problem 7-12. If your answers are incorrect, review Section 7-6 (Study Exercise 7-15) in your text.

Proceed to Problem 7-12.

PROBLEM 7-12

Write the correct formula for each of the following compounds:

(a) milk of magnesia   (a) _____

(b) dry ice   (b) _____

(c) baking soda   (c) _____

(d) ammonia   (d) _____

(e) vinegar   (e) _____

ANSWERS TO PROBLEM 7-12

(a) $Mg(OH)_2$

(b) CO_2

(c) $NaHCO_3$

(d) NH_3

(e) $HC_2H_3O_2$

If your answers are correct, you have completed this chapter. If your answers are incorrect, review Section 7-7 in your text.

These problems conclude the chapter on chemical nomenclature of inorganic compounds.

Now take the sample quiz to see if you have mastered the material in the chapter.

SAMPLE QUIZ

<u>Quiz #7</u> You may use the periodic table.

1. Write the correct name for each of the following compounds:

  (a) KI   _____

  (b) SnS_2 _____

  (c) $FeSO_3$ _____

  (d) $NaC_2H_3O_2$ _____

  (e) $NaClO_4$ _____

  (f) N_2O_5 _____

  (g) $(NH_4)_2SO_4$ _____

  (h) $BaCrO_4$ _____

2. Write the correct formula for each of the following com-
    pounds:

  (a) magnesium nitride   _____

  (b) copper(I) oxide   _____

  (c) ammonium periodate   _____

  (d) milk of magnesia   _____

  (e) sulfur trioxide   _____

  (f) hydrogen sulfide   _____

  (g) nitric acid   _____

  (h) lithium phosphate   _____

3. Classify each of the following compounds as (1) an acid,
    (2) a base, or (3) a salt. Assume that all soluble
    compounds are in aqueous solution.

  (a) K_2SO_4 _____

  (b) $Ca(OH)_2$ _____

  (c) HNO_3 in aqueous   _____
       solution

  (d) $Bi(NO)_3$ _____

101

1. (a) potassium iodide; (b) tin(IV) sulfide or stannic sulfide; (c) iron(II) sulfite or ferrous sulfite; (d) sodium acetate; (e) sodium perchlorate; (f) dinitrogen pentoxide; (g) ammonium sulfate; (h) barium chromate

2. (a) Mg_3N_2; (b) Cu_2O; (c) NH_4IO_4; (d) $Mg(OH)_2$; (e) SO_3;

   (f) H_2S; (g) HNO_3; (h) Li_3PO_4

3. (a) (3) salt; (b) (2) base; (c) (1) acid; (d) (3) salt

CHAPTER 8

CALCULATIONS INVOLVING ELEMENTS AND COMPOUNDS

In Chapter 8 we will discuss quantitative (how much?) calculations involving elements and compounds. We will use
dimensional analysis to solve these problems.

SELECTED TOPICS

1. In the formula of a compound, the subscripts represent the
   number of atoms of the respective elements in a molecule or
   formula unit of the compound. Using this information and
   the approximate atomic masses of the elements (see inside
   back cover of your text), the formula mass or molecular
   mass of a compound can be calculated. The formula mass is
   used for compounds that are written as formula units and
   that have primarily ionic bonding. Molecular mass is used
   for compounds that exist as molecules and that have primarily covalent bonds. The methods for calculating formula
   and molecular masses are the same. The formula for methane
   gas is CH_4 and consists of one atom of carbon and four
   atoms of hydrogen. Using the table of approximate atomic
   masses found inside the back cover of your text, the molecular mass of CH_4 is 16.0 amu (1 x 12.0 amu for C + 4 x 1.0
   amu for H = 12.0 amu + 4.0 amu = 16.0 amu for CH_4). See
   Section 8-1 in your text.

2. A mole (abbreviated mol) is the amount of a substance containing the same number of atoms, formula units, molecules

   or ions as there are atoms in exactly 12 g of ^{12}C atoms. In

   exactly 12 g of ^{12}C atoms, there are 6.02×10^{23} atoms

   (Avogadro's number). Therefore, in one mole of ^{12}C atoms

   there are 6.02×10^{23} atoms; this number of atoms has a

   mass of exactly 12 g, the atomic mass for ^{12}C expressed in
   grams. This statement can be expanded for atoms of any
   element, formula units or molecules of a compound, and ions
   in a compound as follows:

   1 mol of atoms of an element = 6.02×10^{23} atoms of the
        element = atomic mass of the element in grams

   1 mol of a compound = 6.02×10^{23} formula units or mole-
        cules of the compound = formula or molecular mass
        in grams

   1 mol of ions = 6.02×10^{23} ions = atomic or formula mass
        of the ion in grams

This atomic mass, formula mass, or molecular mass expressed in grams has a special name, the molar mass. The molar mass is the mass in grams of one mole of any substance, element or compound.

The subscripts in the formula of a compound can also represent the number of moles of ions or atoms of the elements in one mole of molecules or formula units of the compound. For example, one mole of CH_4, contains one mole of carbon atoms and four moles of hydrogen atoms. See Section 8-2 in your text.

3. Experiments have shown that for any GAS, 6.02×10^{23} molecules of the gas or 1 mol of gas molecules occupies a volume of 22.4 L at $0^{\circ}C$ (273 K) and a pressure of 760 mm Hg (torr) - standard temperature and pressure (STP). This volume of 22.4 L occupied by 1 mol of molecules of any GAS at $0^{\circ}C$ and 760 mm Hg is called the molar volume of a gas. Using the molar volume of a gas we can make the following calculations:

   (1) moles or mass of a gas in any volume of the gas at STP
   (2) molecular mass and molar mass of a gas by solving for grams per mole of the gas which is numerically equal to the molecular mass in amu and the molar mass in grams
   (3) the density of a gas by solving for grams per liter of the gas at STP
   See Section 8-3 in your text.

4. The percent composition of each element in a compound can be calculated from the formula of the compound and the approximate atomic mass of the elements or from the experimental analysis of the compound. For example, methane gas (CH_4) has a percent composition of 75.0 percent carbon ($\frac{12.0 \ \text{amu}}{16.0 \ \text{amu}} \times 100 = 75.0 \ \% \ C$) and 25.0 percent hydrogen ($\frac{4.0 \ \text{amu}}{16.0 \ \text{amu}} \times 100 = 25.0 \ \% \ H$).
   See Section 8-4 in your text.

5. The empirical formula is the simplest formula of a compound which contains the smallest whole-number ratio of the atoms that are present. The empirical formula is determined from the percent composition of the compound as determined experimentally from analysis of the compound in the laboratory and the approximate atomic masses of the elements. In the empirical formula the ratio of the atoms is given in the smallest whole numbers possible. The molecular formula is the actual number of atoms of each element present in one molecule of the compound. It is determined from the empirical formula and the molecular mass of the compound as determined experimentally by various laboratory methods. For example, the empirical formula

for benzene, an organic compound, is found to be CH; but its molecular formula is C_6H_6 using the molecular mass determined by experimentation. See Section 8-5 in your text.

PROBLEM 8-1

Calculate the formula or molecular mass of each of the following compounds. (Use the table of approximate atomic masses in the inside back cover of your text.)

(a) $C_{12}H_{22}O_{11}$ (sucrose, table sugar)   (a) _____

(b) $Ca(CN)_2$ (b) _____

(c) K_2HPO_4 (c) _____

(d) $Fe_3(PO_4)_2$ (d) _____

ANSWERS AND SOLUTIONS TO PROBLEM 8-1

(a) 342 amu

$$12 \times 12.0 = 144 \text{ amu}$$
$$22 \times 1.0 = 22 \text{ amu}$$
$$11 \times 16.0 = \underline{176} \text{ amu}$$
$$\overline{342} \text{ amu}$$

(b) 92.1 amu

$$1 \times 40.1 = 40.1 \text{ amu}$$
$$2 \times 12.0 = 24.0 \text{ amu} \quad \text{clear ( ), so 2 C atoms}$$
$$2 \times 14.0 = \underline{28.0} \text{ amu} \quad \text{clear ( ), so 2 N atoms}$$
$$\overline{92.1} \text{ amu}$$

(c) 174.2 amu

$$2 \times 39.1 = 78.2 \text{ amu}$$
$$1 \times 1.0 = 1.0 \text{ amu}$$
$$1 \times 31.0 = 31.0 \text{ amu}$$
$$4 \times 16.0 = \underline{64.0} \text{ amu}$$
$$\overline{174.2} \text{ amu}$$

(d) 357 amu

$$3 \times 55.8 = 167 \text{ amu (3 significant digits)}$$
$$2 \times 31.0 = 62.0 \text{ amu}$$
$$8 \times 16.0 = \underline{128} \text{ amu (3 significant digits)}$$
$$\overline{357} \text{ amu}$$

If your answers are correct, proceed to Problem 8-2. If your answers are incorrect, review Section 8-1 (Examples 8-1 and 8-2 and Study Exercise 8-1) in your text.

Proceed to Problem 8-2.

PROBLEM 8-2

Calculate the number of

(a) moles of sucrose ($C_{12}H_{22}O_{11}$) in 187 g sucrose

(a) _____

(b) moles of sucrose ($C_{12}H_{22}O_{11}$) in 9.83×10^{23} molecules of sucrose

(b) _____

(c) moles of calcium cyanide in $32\overline{0}$ g calcium cyanide

(c) _____

106

(d) moles of calcium cyanide in 6.84×10^{24} formula units of calcium cyanide

(d) _____

(e) moles of cyanide ions in 2.75 mol of calcium cyanide

(e) _____

ANSWERS AND SOLUTIONS TO PROBLEM 8-2

(a) 0.547 mol   187 g̶ ̶C̶$_{12}$H̶$_{22}$O̶$_{11}$ $\times$ $\dfrac{1 \text{ mol } C_{12}H_{22}O_{11}}{342 \text{ g̶ ̶C̶}_{12}H̶_{22}O̶_{11}}$ =

        0.547 mol $C_{12}H_{22}O_{11}$

        Molar mass = 342 g, see Problem 8-1

(b) 1.63 mol 9.83×10^{23} m̶o̶l̶e̶c̶u̶l̶e̶s̶ ̶C̶$_{12}$H̶$_{22}$O̶$_{11}$ $\times$

        $\dfrac{1 \text{ mol } C_{12}H_{22}O_{11}}{6.02 \times 10^{23} \text{ m̶o̶l̶e̶c̶u̶l̶e̶s̶ ̶C̶}_{12}H̶_{22}O̶_{11}}$ =

        1.63 mol $C_{12}H_{22}O_{11}$

        In 1 mol of anything there are 6.02×10^{23} units (Avogadro's number).

(c) 3.47 mol $32\overline{0}$ g̶ ̶C̶a̶(̶C̶N̶)̶$_2$ $\times$ $\dfrac{1 \text{ mol } Ca(CN)_2}{92.1 \text{ g̶ ̶C̶a̶(̶C̶N̶)̶}_2}$ = 3.47 mol $Ca(CN)_2$

        Molar mass - 92.1 g, see Problem 8-1

(d) 11.4 mol 6.84×10^{24} f̶o̶r̶m̶u̶l̶a̶ ̶u̶n̶i̶t̶s̶ ̶C̶a̶(̶C̶N̶)̶$_2$ $\times$

        $\dfrac{1 \text{ mol } Ca(CN)_2}{6.02 \times 10^{23} \text{ f̶o̶r̶m̶u̶l̶a̶ ̶u̶n̶i̶t̶s̶ ̶C̶a̶(̶C̶N̶)̶}_2}$ =

        $1.14 \times 10^{24-23}$ mol = 1.14×10 = 11.4 mol $Ca(CN)_2$

(e) 5.50 mol   2.75 m̶o̶l̶ ̶C̶a̶(̶C̶N̶)̶$_2$ $\times$ $\dfrac{2 \text{ mol } CN^-}{1 \text{ m̶o̶l̶ ̶C̶a̶(̶C̶N̶)̶}_2}$ =

        5.50 mol CN^- ions

        In 1 mol $Ca(CN)_2$ there are 2 mol of CN^- ions.

If your answers are correct, proceed to Problem 8-3. If your answers are incorrect, review Section 8-2 (Examples 8-3, 8-4, 8-5, 8-6, 8-7, and 8-8 and Study Exercises 8-2, 8-3, and 8-4) in your text.

Proceed to Problem 8-3.

PROBLEM 8-3

(a) Given 0.654 mol of sucrose ($C_{12}H_{22}O_{11}$), calculate (1) the number of grams, and (2) the number of molecules of sucrose.

(1) _____

(2) _____

(b) Given 0.362 mol of calcium cyanide, calculate (1) the number of grams of calcium cyanide, (2) the number of formula units of calcium cyanide and (3) the number of cyanide ions.

(1) _____

(2) _____

(3) _____

ANSWERS AND SOLUTIONS TO PROBLEM 8-3

(a) (1) 224 g   (1) 0.654 mol $C_{12}H_{22}O_{11}$ x $\dfrac{342 \text{ g } C_{12}H_{22}O_{11}}{1 \text{ mol } C_{12}H_{22}O_{11}}$

= 224 g $C_{12}H_{22}O_{11}$

(2) 3.94×10^{23} molecules

(2) 0.654 mol C₁₂H₂₂O₁₁ x

$$\frac{6.02 \times 10^{23} \text{ molecules } C_{12}H_{22}O_{11}}{1 \text{ mol } C_{12}H_{22}O_{11}}$$

$$= 3.94 \times 10^{23} \text{ molecules } C_{12}H_{22}O_{11}$$

(b) (1) 33.3 g   (1) 0.362 mol Ca(CN)₂ x $\dfrac{92.1 \text{ g } Ca(CN)_2}{1 \text{ mol } Ca(CN)_2}$

$$= 33.3 \text{ g } Ca(CN)_2$$

(2) 2.18×10^{23} formula units

(2) 0.362 mol Ca(CN)₂ x

$$\frac{6.02 \times 10^{23} \text{ formula units } Ca(CN)_2}{1 \text{ mol } Ca(CN)_2}$$

$$= 2.18 \times 10^{23} \text{ formula units } Ca(CN)_2$$

(3) 4.36×10^{23} CN⁻ ions

(3) 0.362 mol Ca(CN)₂ x $\dfrac{2 \text{ mol } CN^-}{1 \text{ mol } Ca(CN)_2}$ x

$$\frac{6.02 \times 10^{23} \text{ CN}^- \text{ ions}}{1 \text{ mol } CN^-} = 4.36 \times 10^{23} \text{ CN}^- \text{ ions}$$

If your answers are correct, proceed to Problem 8-4. If your
answers are incorrect, review Section 8-2 (Examples 8-9, 8-10,
8-11, 8-12, and 8-13 and Study Exercises 8-5, 8-6, 8-7, and
8-8) in your text.

Proceed to Problem 8-4.

PROBLEM 8-4

Calculate to three significant digits the mass in grams of one
atom of

(a) an isotope of Si, atomic mass = 28.0 amu

(a) _____

(b) an isotope of Kr, atomic mass = 81.9 amu

(b) _____

109

ANSWERS AND SOLUTIONS TO PROBLEM 8-4

(a) 4.65×10^{-23} g Si

Using the atomic mass units of Si (28.0 amu), as
$\dfrac{28.0 \text{ g Si}}{1 \text{ mol Si atoms}}$ and Avogadro's number,

$\dfrac{6.02 \times 10^{23} \text{ Si atoms}}{1 \text{ mol Si atoms}}$, calculate the mass in grams

of one atom of Si as follows:

$$1 \text{ atom Si} \times \frac{1 \text{ mol Si atoms}}{6.02 \times 10^{23} \text{ atoms Si}} \times \frac{28.0 \text{ g Si}}{1 \text{ mol Si atoms}}$$

$$= 4.65 \times 10^{-23} \text{ g Si}$$

(b) 1.36×10^{-22} g Kr

$$1 \text{ atom Kr} \times \frac{1 \text{ mol Kr atoms}}{6.02 \times 10^{23} \text{ atoms Kr}} \times \frac{81.9 \text{ g Kr}}{1 \text{ mol Kr atoms}}$$

$$= 13.6 \times 10^{-23} \text{ g Kr} = 1.36 \times 10^{-22} \text{ g Kr}$$

If your answers are correct, proceed to Problem 8-5. If your answers are incorrect, review Section 8-2 (Example 8-14 and Study Exercise 8-9) in your text.

Proceed to Problem 8-5.

PROBLEM 8-5

Given 9.75 liters of nitrogen gas molecules at STP, calculate

(a)   the number of moles of nitrogen gas molecules

(a) _____

(b)   the number of grams of nitrogen gas molecules

(b) _____

ANSWERS AND SOLUTIONS TO PROBLEM 8-5

(a)   0.435 mol   9.75 ~~L N$_2$ STP~~ x $\dfrac{1 \text{ mol N}_2}{22.4 \text{ L N}_2 \text{ STP}}$ = 0.435 mol N$_2$

One mole of gas molecules occupies a volume of 22.4 L at STP.

(b)   12.2 g   9.75 ~~L N$_2$ STP~~ x $\dfrac{1 \text{ mol N}_2}{22.4 \text{ L N}_2 \text{ STP}}$ x $\dfrac{28.0 \text{ g N}_2}{1 \text{ mol N}_2}$

= 12.2 g N$_2$

Nitrogen is a diatomic molecule (N$_2$). The molar mass of N$_2$ is 28.0 g (2 x 14.0 $=$ 28.0 amu).

If your answers are correct, proceed to Problem 8-6. If your answers are incorrect, review Section 8-3 (Examples 8-15 and 8-16 and Study Exercise 8-10) in your text.

Proceed to Problem 8-6.

PROBLEM 8-6

Calculate the molecular mass and molar mass of a gas given the following data:

(a)   8.90 g of the gas occupies a volume of 4.25 L at STP

(a) _____

(b)   the density of the gas is 1.34 g/L at STP

(b) _____

ANSWERS AND SOLUTIONS TO PROBLEM 8-6

(a)   46.9 amu,   To calculate the molecular mass and molar mass,
      46.9 g   we solve for the units grams per mole (g/mol)
                  of the gas which is numerically equal to the
                  molecular mass in amu and the molar mass in
                  grams.

$$\frac{8.90 \text{ g}}{4.25 \text{ L STP}} \times \frac{22.4 \text{ L STP}}{1 \text{ mol}} = 46.9 \frac{\text{g}}{\text{mol}} \text{ , } 46.9 \text{ amu, } 46.9 \text{ g}$$

(b) 30.0 amu, $\frac{1.34 \text{ g}}{1 \text{ L STP}} \times \frac{22.4 \text{ L STP}}{1 \text{ mol}} = 30.0 \frac{\text{g}}{\text{mol}}$, 30.0 amu, 30.0 g
30.0 g

If your answers are correct, proceed to Problem 8-7. If your answers are incorrect, review Section 8-3 (Example 8-17 and 8-18 and Study Exercise 8-11) in your text.

Proceed to Problem 8-7.

PROBLEM 8-7

Calculate

(a) the density of carbon dioxide gas at STP

(a) _____

(b) the volume in liters occupied by 6.85 g of carbon dioxide gas at STP

(b) _____

ANSWERS AND SOLUTIONS TO PROBLEM 8-7

(a) 1.96 g/L The units of density for a gas are g/L.

$$\frac{44.0 \text{ g } CO_2}{1 \text{ mol } CO_2} \times \frac{1 \text{ mol } CO_2}{22.4 \text{ L STP}} = 1.96 \frac{\text{g}}{\text{L at STP}}$$

The formula for carbon dioxide is CO_2. Carbon dioxide has a molar mass of 44.0 g (1 x 12.0 amu + 2 x 16.0 amu = 12.0 amu + 32.0 amu = 44.0 amu).

(b) 3.49 L $6.85 \text{ g } CO_2 \times \frac{1 \text{ mol } CO_2}{44.0 \text{ g } CO_2} \times \frac{22.4 \text{ L } CO_2 \text{ STP}}{1 \text{ mol } CO_2}$
= 3.49 L CO_2 at STP

If your answers are correct, proceed to Problem 8-8. If your answers are incorrect, review Section 8-3 (Example 8-19 and 8-20 and Study Exercises 8-12 and 8-13) in your text.

Proceed to Problem 8-8.

PROBLEM 8-8

Calculate

(a) the percent composition of glucose ($C_6H_{12}O_6$)

(a) ______________________

(b) the percent of metal in a metal oxide if 2.50 g of the metal combined with 1.50 g of oxygen

(b) ______________________

(c) the number of grams of sodium ions in 93.6 g of sodium phosphate (Na_3PO_4)

(c) ______________________

ANSWERS AND SOLUTIONS TO PROBLEM 8-8

(a) 40.0 % C
6.7 % H
53.3 % O

Calculate the molecular mass of $C_6H_{12}O_6$ as follows:

$$\begin{aligned} 6 \times 12.0 &= 72.0 \text{ amu} \\ 12 \times 1.0 &= 12 \text{ amu} \\ 6 \times 16.0 &= \underline{96.0 \text{ amu}} \\ & 180 \text{ amu} \end{aligned}$$

Calculate the percent of each element in the compound as follows:

$$\frac{72.0 \cancel{\text{ amu}}}{180 \cancel{\text{ amu}}} \times 100 = 40.0 \% \text{ C};$$

$$\frac{12 \cancel{\text{ amu}}}{180 \cancel{\text{ amu}}} \times 100 = 6.7 \% \text{ H}; \quad \frac{96.0 \cancel{\text{ amu}}}{180 \cancel{\text{ amu}}} \times 100 = 53.3 \% \text{ O}$$

(b) 62.5 %   The total mass of the oxide is 4.00 g (2.50 g metal + 1.50 g oxygen).

$$\frac{2.50 \text{ g metal}}{4.00 \text{ g oxide}} \times 100 = 62.5 \text{ %}$$

(c) 39.4 g   Calculate the formula mass of Na_3PO_4 as follows:

$$
\begin{array}{rcl}
3 \times 23.0 & = & 69.0 \text{ amu} \\
1 \times 31.0 & = & 31.0 \text{ amu} \\
4 \times 16.0 & = & \underline{64.0 \text{ amu}} \\
& & 164.0 \text{ amu}
\end{array}
$$

In 164.0 g Na_3PO_4, there are 69.0 g Na; therefore, calculate the number of grams of Na in 93.6 g Na_3PO_4 as follows:

$$93.6 \text{ g } Na_3PO_4 \times \frac{69.0 \text{ g Na}}{164.0 \text{ g } Na_3PO_4} = 39.4 \text{ g Na}$$

If your answers are correct, proceed to Problem 8-9. If your answers are incorrect, review Section 8-4 (Examples 8-21, 8-22, 8-23, and 8-24 and Study Exercises 8-14, 8-15, and 8-16) in your text.

Proceed to Problem 8-9.

PROBLEM 8-9

Calculate the empirical formula for each of the following compounds:

(a) 30.4 percent nitrogen and 69.6 percent oxygen

(a) _____

(b) 47.3 percent potassium, 18.8 percent phosphorus, and 33.9 percent oxygen

(b) _____

(c) 0.750 g sulfur combines with 1.12 g oxygen

(c) _____

114

ANSWERS AND SOLUTIONS TO PROBLEM 8-9

(a) NO_2 In exactly $1\overline{00}$ g of the compound there are 30.4 g N and 69.6 g O.

    (1) Moles of atoms of each element

$$30.4 \text{ g N} \times \frac{1 \text{ mol N atoms}}{14.0 \text{ g N}} = 2.17 \text{ mol N atoms}$$

$$69.6 \text{ g O} \times \frac{1 \text{ mol O atoms}}{16.0 \text{ g O}} = 4.35 \text{ mol O atoms}$$

    (2) Divide by the smallest

$$\frac{2.17}{2.17} = 1 \text{ for N} \qquad \frac{4.35}{2.17} \cong 2 \text{ for O}$$

    The empirical formula is NO_2

(b) $K_4P_2O_7$ In exactly $1\overline{00}$ g of the compound there are 47.3 g K, 18.8 g P, and 33.9 g O.

    (1) Moles of atoms of each element

$$47.3 \text{ g K} \times \frac{1 \text{ mol K atoms}}{39.1 \text{ g K}} = 1.21 \text{ mol K atoms}$$

$$18.8 \text{ g P} \times \frac{1 \text{ mol P atoms}}{31.0 \text{ g P}} = 0.606 \text{ mol P atoms}$$

$$33.9 \text{ g O} \times \frac{1 \text{ mol O atoms}}{16.0 \text{ g O}} = 2.12 \text{ mol O atoms}$$

    (2) Divide by the smallest

$$\frac{1.21}{0.606} \cong 2 \text{ for K} \qquad \frac{0.606}{0.606} = 1 \text{ for P} \qquad \frac{2.12}{0.606} \cong 3.5 \text{ for O}$$

    Convert these relative ratios to small whole numbers by multiplying by 2; the empirical formula is $K_4P_2O_7$.

(c) SO_3 (1) Moles of atoms of each element

$$0.750 \text{ g S} \times \frac{1 \text{ mol S atoms}}{32.1 \text{ g S}} = 0.0234 \text{ mol S atoms}$$

$$1.12 \text{ g O} \times \frac{1 \text{ mol O atoms}}{16.0 \text{ g O}} = 0.0700 \text{ mol O atoms}$$

(2) Divide by the smallest

$$\frac{0.0234}{0.0234} = 1 \text{ for S} \qquad \frac{0.0700}{0.0234} \cong 3 \text{ for O}$$

The empirical formula is SO_3.

If your answers are correct, proceed to Problem 8-10. If your answers are incorrect, review Section 8-5 (Examples 8-25 and 8-26 and Study Exercise 8-17) in your text.

Proceed to Problem 8-10.

PROBLEM 8-10

Determine the molecular formula for the following compounds from the following experimental data:

(a) 85.6 percent carbon, 14.4 percent hydrogen, and molecular mass of 56.1 amu

(a) _____

(b) 80.0 percent carbon, 20.0 percent hydrogen, and a density of 1.34 g/L at STP

(b) _____

ANSWERS AND SOLUTIONS TO PROBLEM 8-10

(a) C_4H_8 Empirical Formula

In exactly $10\overline{0}$ g of the compound there are 85.6 g C and 14.4 g H.

(1) Moles of atoms of each element

$$85.6 \text{ g C} \times \frac{1 \text{ mol C atoms}}{12.0 \text{ g C}} = 7.13 \text{ mol C atoms}$$

$$14.4 \text{ g H} \times \frac{1 \text{ mol H atoms}}{1.0 \text{ g H}} = 14.4 \text{ mol H atoms}$$

(2) Divide by smallest

$$\frac{7.13}{7.13} = 1 \text{ for C} \qquad \frac{14.4}{7.13} \cong 2 \text{ for H}$$

Empirical formula is CH_2.

Molecular Formula

Calculate the empirical formula mass as follows:

$$1 \times 12.0 = 12.0 \text{ amu}$$
$$2 \times 1.0 = \underline{2.0 \text{ amu}}$$
$$14.0 \text{ amu}$$

$$\frac{\text{molecular mass}}{\text{empirical formula mass}} = \frac{56.1 \text{ amu}}{14.0 \text{ amu}} = 4.01 \text{ or}$$

approximately 4

Molecular formula is $(CH_2)_4 = C_4H_8$

(b) C_2H_6 Empirical Formula

In exactly $1\overline{00}$ g of the compound there are 80.0 g C and 20.0 g H.

(1) Moles of atoms of each element

$$80.0 \text{ g C} \times \frac{1 \text{ mol C atoms}}{12.0 \text{ g C}} = 6.67 \text{ mol C atoms}$$

$$20.0 \text{ g H} \times \frac{1 \text{ mol H atoms}}{1.0 \text{ g H}} = 2\overline{0} \text{ mol H atoms}$$

(2) Divide by the smallest

$$\frac{6.67}{6.67} = 1 \text{ for C} \qquad \frac{2\overline{0}}{6.67} \cong 3 \text{ for H}$$

Empirical formula is CH_3.

Molecular Formula

Calculate the molecular mass as follows:

$$1.34 \frac{\text{g}}{\text{L STP}} \times 22.4 \frac{\text{L STP}}{1 \text{ mol}} = 30.0 \frac{\text{g}}{\text{mol}}, \ 30.0 \text{ amu}$$

117

Calculate the empirical formula mass as follows:

$$1 \times 12.0 = 12.0 \text{ amu}$$
$$3 \times 1.0 = \underline{3.0 \text{ amu}}$$
$$15.0 \text{ amu}$$

$$\frac{\text{molecular mass}}{\text{empirical formula mass}} = \frac{30.0 \cancel{\text{amu}}}{15.0 \cancel{\text{amu}}} = 2$$

Molecular formula is $(CH_3)_2 = C_2H_6$.

If your answers are correct, you have completed this chapter. If your answers are incorrect, review Section 8-5 (Problem Examples 8-27, 8-28, and 8-29 and Study Exercise 8-18) in your text.

These problems conclude the chapter on calculations involving elements and compounds.

Now take some sample quizzes to see if you have mastered the material in this chapter.

SAMPLE QUIZZES

The quizzes divide this chapter into two parts: Quiz #8A (Sections 8-1 to 8-3) and Quiz #8B (Sections 8-4 to 8-5).

Quiz #8A (Sections 8-1 to 8-3)

Element	Atomic Mass (amu)
C	12.0
O	16.0

1. Calculate the number of moles in 36.0 g of carbon dioxide.

2. Calculate the number of grams in 0.574 mol of carbon dioxide.

3. Calculate the number of carbon dioxide molecules in 32.5 g of carbon dioxide.

4. Calculate the number of liters of carbon dioxide (at STP) in 41.0 g of carbon dioxide.

5. Calculate the molecular mass and molar mass of a gas which has a density of 1.20 g/L at STP.

Solutions and Answers for Quiz #8A

1. $36.0 \text{ g CO}_2 \times \dfrac{1 \text{ mol CO}_2}{44.0 \text{ g CO}_2} = 0.818 \text{ mol CO}_2$

2. $0.574 \text{ mol CO}_2 \times \dfrac{44.0 \text{ g CO}_2}{1 \text{ mol CO}_2} = 25.3 \text{ g CO}_2$

3. $32.5 \text{ g CO}_2 \times \dfrac{1 \text{ mol CO}_2}{44.0 \text{ g CO}_2} \times \dfrac{6.02 \times 10^{23} \text{ molecules}}{1 \text{ mol CO}_2}$

   $= 4.45 \times 10^{23}$ molecules

4. $41.0 \text{ g CO}_2 \times \dfrac{1 \text{ mol CO}_2}{44.0 \text{ g CO}_2} \times \dfrac{22.4 \text{ L STP}}{1 \text{ mol CO}_2} = 20.9 \text{ L STP}$

5. $1.20 \dfrac{\text{g}}{\text{L STP}} \times \dfrac{22.4 \text{ L STP}}{1 \text{ mol}} = 26.9 \dfrac{\text{g}}{\text{mol}}$, 26.9 amu, 26.9 g

Quiz #8B (Sections 8-4 to 8-5)

Element	Atomic Mass (amu)
C	12.0
H	1.0
O	16.0
N	14.0

1. Calculate the percent composition of the food preservative benzoic acid ($C_7H_6O_2$).

2. Calculate the number of grams of oxygen in 32.0 g of ethyl alcohol (C_2H_6O).

3. Determine the empirical formula for a compound that gave on analysis 27.3 percent carbon and 72.7 percent oxygen.

4. Determine the molecular formula for a compound with the following composition: 92.2 percent carbon and 7.8 percent hydrogen. Experiment found that 4.00 g of the gas occupies 3.45 liters at STP.

Solutions and Answers for Quiz #8B

1. 7 x 12.0 =   84.0 amu
    6 x 1.0 =   6.0 amu
    2 x 16.0 =   32.0 amu
                 122.0 amu

    $\frac{84.0 \text{ amu}}{122.0 \text{ amu}}$ x 100 = 68.9 % C $\frac{32.0 \text{ amu}}{122.0 \text{ amu}}$ x 100 = 26.2 % O

    $\frac{6.0 \text{ amu}}{122.0 \text{ amu}}$ x 100 = 4.9 % H

2. 2 x 12.0 = 24.0 amu
    6 x 1.0 = 6.0 amu
    1 x 16.0 = 16.0 amu
               46.0 amu

    32.0 g C_2H_6O x $\frac{16.0 \text{ g O}}{46.0 \text{ g } C_2H_6O}$ = 11.1 g O

3. $27.3 \cancel{g\ C} \times \dfrac{1\ \text{mol C atoms}}{12.0\ \cancel{g\ C}} = 2.28\ \text{mol C atoms} \qquad \dfrac{2.28}{2.28} = 1\ \text{for C}$

$72.7 \cancel{g\ O} \times \dfrac{1\ \text{mol O atoms}}{16.0\ \cancel{g\ O}} = 4.54\ \text{mol O atoms} \qquad \dfrac{4.54}{2.28} \cong 2\ \text{for O}$

Empirical formula is CO_2

4. <u>Empirical Formula</u>

$92.2 \cancel{g\ C} \times \dfrac{1\ \text{mol C atoms}}{12.0\ \cancel{g\ C}} = 7.68\ \text{mol C atoms} \qquad \dfrac{7.8}{7.68} \cong 1\ \text{for C}$

$7.8 \cancel{g\ H} \times \dfrac{1\ \text{mol H atoms}}{1.0\ \cancel{g\ H}} = 7.8\ \text{mol H atoms} \qquad \dfrac{7.68}{7.68} = 1\ \text{for H}$

Empirical formula is CH.

<u>Molecular Formula</u>

$\dfrac{4.00\ g}{3.45\ \cancel{L\ STP}} \times \dfrac{22.4\ \cancel{L\ STP}}{1\ \text{mol}} = 26.0\ \dfrac{g}{\text{mol}},\ 26.0\ \text{amu} \qquad \dfrac{26.0\ \cancel{amu}}{13.0\ \cancel{amu}} = 2$

Molecular formula is $(CH)_2 = C_2H_2$

CHAPTER 9

CHEMICAL EQUATIONS

In chapter 9 we will consider (1) balancing chemical equations, and (2) completing and balancing chemical equations. By using chemical equations we will be able to understand the chemical properties and chemical changes of elements and compounds.

SELECTED TOPICS

1. A chemical equation is a shorthand method for expressing a chemical reaction in symbols and formulas. A chemical equation must be balanced. We balance chemical equations because of the law of conservation of mass. This law requires that the number of atoms or moles of atoms of each element be the same on both sides of the equation. See Section 9-1 in your text.

2. In a chemical equation, the substances that combine with one another and change are the reactants and are written on the left. The substances that form and appear are the products and are written on the right. Various terms and symbols are used in chemical equations. These terms, symbols, and their meanings are given in Table 9-1 in your text. See Section 9-2.

3. To balance chemical equations, we follow various guidelines. These guidelines are given in Section 9-3 of your text. One of the most important guidelines is that you must never change a correct formula of a compound to balance the equation. A summary of these guidelines is as follows:

    1. Write the correct formulas.
    2. Start with a specific element in a compound with the most atoms.
    3. Balance the polyatomic ions.
    4. Balance the H atoms and then the O atoms.
    5. Check the coefficients to see that they are whole numbers and in the lowest possible ratio.
    6. Check each atom or polyatomic ion with a ✓.

    Word equations are another form of chemical equations without coefficients. A word equation expresses the chemical equation in words instead of symbols and formulas. To write chemical equations from word equations you must write the correct formulas for the elements or compounds from the names (see Chapter 7, Chemical Nomenclature of Inorganic Compounds). See Section 9-3 in your text.

4. There are five basic types of simple chemical reactions for which we can write chemical equations:

    1. combination reactions
    2. decomposition reactions
    3. single-replacement reactions
    4. double-replacement reactions
    5. neutralization reactions

    You will write the products for single-replacement, double-replacement, and neutralization reactions. For combination and decomposition reactions, the products will be given. You will need to balance and identify all five types of reactions. See Section 9-5 in your text.

5. In <u>combination</u> <u>reactions</u>, two or more substances (either elements or compounds) react to produce one substance (always a compound). The following general equation represents this reaction:

    A + Z ---> AZ where A and Z are elements or compounds.

    Some combination reactions involving oxygen gas are called combustion reactions. Section 9-6 in your text.

6. In <u>decomposition</u> <u>reactions</u>, one substance (always a compound) breaks down to form two or more substances (either element or compound). The following general equation represents this reaction:

    AZ ---> A + Z where A and Z are elements or compounds.

    Heat may be necessary for the reaction to go. See Section 9-7 in your text.

7. In <u>single-replacement</u> <u>reactions</u>, an element and a compound react and the element replaces another element in the compound. The following two general equations represent these reactions:

    (1) a metal replacing a metal ion in its salt or acid

         A + BZ ---> AZ + B

    (2) a nonmetal replacing a nonmetal ion in its salt or acid

         X + BZ ---> BX + Z

    In the first general equation, the replacement depends on the two metals involved. These metals are arranged in a series called the <u>electromotive</u> or <u>activity series</u>. An element high in the series will displace an element lower in the series from an aqueous solution of its salt, or hydrogen from an acid if the element is above hydrogen in

123

the series. This series is given in Section 9-8 and on the inside back cover of your text. Unless instructed otherwise, you need not memorize the order of this series as it will be given to you on quizzes and exams. You must, however, be able to interpret and use this order.

In the second general equation, the replacement depends on the two nonmetals involved. A series similar to the electromotive or activity series exists for the halogens, F_2,

Cl_2, Br_2, and I_2. Fluorine will replace chloride ions from

an aqueous solution of a chloride salt or acid; chlorine will replace bromide; and bromine will replace iodide. Notice that this halogen series follows the nonmetallic properties given in the periodic table. Since the periodic table will be given to you on quizzes and exams, you need not memorize the order. See Section 9-8 in your text.

8. In <u>double replacement reactions</u>, two compounds react and the cation of one compound exchanges places with the cation of the other compound. The following general equation represents this reaction:

    AX + BZ ---> AZ + BX

Many double-replacement reactions generally occur (1) when a precipitate is formed, (2) when a weakly ionized species is produced as a product, such as water, or (3) when a gas is formed as a product. To determine whether or not a precipitate is formed, we have rules for the solubility of inorganic substances in water. These rules are given in Section 9-9 and on the inside back cover of your text. Unless instructed otherwise, you need not memorize these rules as they will be given to you on quizzes and exams. You must be able to interpret and use the rules. See Section 9-9 in your text.

9. In <u>neutralization reactions</u> an acid or an acid oxide reacts with a base or basic oxide usually producing water as one of the products. The following general equation represents this reaction:

    HX + MOH ---> MX + HOH

where HX is an acid and MOH is a base. Water is one of the products. See Section 9-10 in your text.

PROBLEM 9-1

Balance each of the following equations by inspection:

(a) $Fe(s) + Cl_2(g) \longrightarrow FeCl_3(s)$

(b) $Sn(s) + HCl(aq) \longrightarrow SnCl_2(aq) + H_2(g)$

(c) $Fe(OH)_3(s) + H_2SO_4(aq) \longrightarrow Fe_2(SO_4)_3(aq) + H_2O(\ell)$

ANSWERS AND SOLUTIONS TO PROBLEM 9-1

(a) $2\ Fe(s) + 3\ Cl_2(g) \longrightarrow 2\ FeCl_3(s)$

Begin with the compound with the greatest number of atoms, $FeCl_3$. To balance the chlorine atoms, place a 2 in front of the $FeCl_3$ and a 3 in front of the Cl_2. Do NOT change the formulas of the compounds to balance the equation. Balance the Fe atoms with a 2 in front of the Fe. Check each atom.

(b) $Sn(s) + 2\ HCl(aq) \longrightarrow SnCl_2(aq) + H_2(g)$

Begin with $SnCl_2$. Balance the Cl atoms by placing a 2 in front of HCl. This balances the hydrogen atoms. Check each atom.

(c) $2\ Fe(OH)_3(s) + 3\ H_2SO_4(aq) \longrightarrow Fe_2(SO_4)_3(aq) + 6\ H_2O(\ell)$

Begin with $Fe_2(SO_4)_3$. Balance the Fe atoms first, by placing a 2 in front of the $Fe(OH)_3$; leave the polyatomic SO_4^{2-} ion until later. Balance the SO_4^{2-} ion by placing a 3 in front of H_2SO_4. In the reactants we now have 12 H atoms [6 in 2 $Fe(OH)_3$ and 6 in 3 H_2SO_4]. Balance the H atoms in the products with a 6 in front of the H_2O to give 6 O atoms in the products and also 6 O atoms in 2 $Fe(OH)_3$.

(The O atoms in the polyatomic SO_4^{2-} ions were previously balanced.) Check each atom.

If your answers are correct, proceed to Problem 9-3. If your answers are incorrect, review Sections 9-3 and 9-4 (Example 9-1 and Study Exercise 9-1) in your text.

Proceed to Problem 9-2.

PROBLEM 9-2

Balance each of the following equations by inspection:

(a) $HCl(\underline{aq}) + Na_2SO_3(\underline{aq}) \longrightarrow NaCl(\underline{aq}) + SO_2(\underline{g}) + H_2O(\ell)$

(b) $P_4O_{10}(\underline{s}) + H_2O(\ell) \longrightarrow H_3PO_4(\underline{aq})$

(c) $C_8H_{18}(\ell) + O_2(\underline{g}) \longrightarrow CO_2(\underline{g}) + H_2O(\underline{g})$

ANSWERS AND SOLUTIONS TO PROBLEM 9-2

(a) $2\ \overset{\checkmark\checkmark}{HCl}(\underline{aq}) + \overset{\checkmark\ \ \checkmark\checkmark}{Na_2SO_3}(\underline{aq}) \longrightarrow 2\ \overset{\checkmark\checkmark}{NaCl}(\underline{aq}) + \overset{\checkmark\checkmark}{SO_2}(\underline{g}) + \overset{\checkmark\checkmark}{H_2O}(\ell)$

Begin with the compound with the greatest number of atoms, Na_2SO_3. Balance the Na atoms by placing a 2 in front of

NaCl. Then balance the Cl atoms with a 2 in front of HCl. The H atoms are balanced. The O atoms are also balanced with 3 in Na_2SO_3 in the reactants and 3 in the products (2 in SO_2 and 1 in H_2O). Check each atom.

(b) $\overset{\checkmark\checkmark}{P_4O_{10}}(\underline{s}) + 6\ \overset{\checkmark\checkmark}{H_2O}(\ell) \longrightarrow 4\ \overset{\checkmark\checkmark\checkmark}{H_3PO_4}(\underline{aq})$

Begin with P_4O_{10}. Balance the P atoms by placing a 4 in

front of H_3PO_4. Balance the H atoms with a 6 in front of

H_2O. This gives 16 O atoms in the reactants (10 in P_4O_{10}

and 6 in 6 H_2O) which are balanced with 16 O atoms in 4

H_3PO_4 in the products. Check each atom.

(c) $C_8H_{18}(\ell) + \dfrac{25}{2}\ O_2(\underline{g}) \overset{\triangle}{\longrightarrow} 8\ CO_2(\underline{g}) + 9\ H_2O(\underline{g})$

Begin with C_8H_{18}. Balance the C atoms by placing an 8 in

front of CO_2. Balance the H atoms with a 9 in front of

water. In the products there are 25 O atoms (16 in 8 CO_2

and 9 in 9 H_2O). To balance the O atoms in the reactants,

place $\dfrac{25}{2}$ in front of O_2. All coefficients must be whole

numbers. To obtain a whole number from $\dfrac{25}{2}$, multiply it by

2 and then multiply all the other coefficients by 2. Check each atom.

126

$$2 \stackrel{\checkmark}{C_8}\stackrel{\checkmark}{H_{18}}(\mathcal{l}) + 25 \stackrel{\checkmark}{O_2}(g) \text{---} > 16 \stackrel{\checkmark}{C}\stackrel{\checkmark}{O_2}(g) + 18 \stackrel{\checkmark}{H_2}\stackrel{\checkmark}{O}(g)$$

If your answers are correct, proceed to Problem 9-3. If your answers are incorrect, re-read Sections 9-3 and 9-4 in your text. Re-work Problems 9-1 and 9-2.

Proceed to Problem 9-3.

PROBLEM 9-3

Change the following word equations into chemical equations and balance them:

(a) aluminum hydroxide + sulfuric acid --->

aluminum sulfate + water

(b) sodium carbonate + hydrochloric acid --->

sodium chloride + carbon dioxide + water

(c) benzene (C_6H_6) + oxygen ---> carbon dioxide + water

ANSWERS AND SOLUTIONS TO PROBLEM 9-3

(a) $2 \stackrel{\checkmark}{Al}(\stackrel{\checkmark}{OH})_3 + 3 \stackrel{\checkmark}{H_2}\stackrel{\checkmark}{SO_4} \text{---} > \stackrel{\checkmark}{Al_2}(\stackrel{\checkmark}{SO_4})_3 + 6 \stackrel{\checkmark}{H_2}\stackrel{\checkmark}{O}$

Write the correct formulas from each of the names of the compounds. See Tables 6-1 (cations), 6-2 (anions), 6-4 (polyatomic ions) and also the periodic table. Once you have written the correct formula, do not change it to balance the equation. Begin with $Al_2(SO_4)_3$. Balance the Al atoms by placing a 2 in front of $Al(OH)_3$. Balance the polyatomic ion $SO_4{}^{2-}$ by placing a 3 in front of H_2SO_4.

Balance the H atoms with a 6 in front of H_2O. This also balances the O atoms. Check each atom.

(b) Na_2CO_3 + 2 HCl ---> 2 NaCl + CO_2 + H_2O

Write the correct formula from each of the names of the compounds. Begin with Na_2CO_3. Balance the Na atoms by placing a 2 in front of NaCl. Balance the Cl atoms by placing a 2 in front of HCl. This balances both the H and O atoms. Check each atom.

(c) C_6H_6 + $\frac{15}{2}$ O_2 $\xrightarrow{\triangle}$ 6 CO_2 + 3 H_2O

Write the correct formulas from each of the names of the compounds. Remember that oxygen is a diatomic molecule. Begin with C_6H_6. Balance the C atoms with a 6 in front of CO_2. Balance the H atoms with a 3 in front of H_2O. This gives 15 O atoms in the products (12 in 6 CO_2 and 3 in 3 H_2O). Place $\frac{15}{2}$ in front of O_2. To obtain whole numbers from $\frac{15}{2}$, multiply it by 2 and then multiply all the other coefficients by 2. Check each atom.

2 C_6H_6 + 15 O_2 $\xrightarrow{\triangle}$ 12 CO_2 + 6 H_2O

If your answers are correct, proceed to Problem 9-4. If your answers are incorrect, review Section 9-4 (Examples 9-2, and 9-3 and Study Exercise 9-2) in your text.

Proceed to Problem 9-4.

PROBLEM 9-4

Balance the following combination reaction equations:

(a) Sr + O_2 $\xrightarrow{\triangle}$ SrO

(b) Na + S $\xrightarrow{\triangle}$ Na_2S

(c) CaO + H_2O ---> Ca(OH)$_2$

ANSWERS AND SOLUTIONS TO PROBLEM 9-4

(a) 2 Sr + O_2 $\xrightarrow{\triangle}$ 2 SrO

(b) 2 Na + S $\xrightarrow{\triangle}$ Na_2S

128

(c) $\overset{\checkmark\checkmark}{CaO} + \overset{\checkmark\ \checkmark}{H_2O} ---> \overset{\checkmark\ \checkmark\checkmark}{Ca(OH)_2}$

A metal oxide plus water forms a base (metal hydroxide).

If your answers are correct, proceed to Problem 9-5. If your answers are incorrect, review Section 9-6 (Study Exercise 9-3) in your text.

Proceed to Problem 9-5.

PROBLEM 9-5

Balance the following decomposition reaction equations:

(a) $KClO_3 \xrightarrow[MnO_2]{\triangle} KCl + O_2$

(b) $SrCO_3 \xrightarrow{\triangle} SrO + CO_2$

(c) $CaCl_2 \cdot 6 H_2O \xrightarrow{\triangle} CaCl_2 + H_2O$

ANSWERS AND SOLUTIONS TO PROBLEM 9-5

(a) $2 \overset{\checkmark\checkmark\checkmark}{KClO_3} \xrightarrow[MnO_2]{\triangle} 2 \overset{\checkmark\checkmark}{KCl} + 3 \overset{\checkmark}{O_2}$

Heating $KClO_3$ in the presence of the catalyst, MnO_2,

decomposes the $KClO_3$ to produce KCl and O_2.

(b) $\overset{\checkmark\checkmark\checkmark}{SrCO_3} \xrightarrow{\triangle} \overset{\checkmark\checkmark}{SrO} + \overset{\checkmark\checkmark}{CO_2}$

Carbonates of Group IIA (2) metals decompose on heating to produce CO_2.

(c) $\overset{\checkmark\ \checkmark}{CaCl_2} \cdot 6 \overset{\checkmark}{H_2O} \xrightarrow{\triangle} \overset{\checkmark\ \checkmark}{CaCl_2} + 6 \overset{\checkmark}{H_2O}$

$CaCl_2 \cdot 6 H_2O$ is a hydrate. Hydrates are crystalline compounds that contain chemically bound water in definite proportions. Hydrates decompose on heating to produce the anhydrous salt (no water) and water. Hydrates will be discussed in detail in chapter 13 (Section 13-7).

If your answers are correct, proceed to Problem 9-6. If your answers are incorrect, review Section 9-7 (Study Exercise 9-4) in your text.

Proceed to Problem 9-6.

PROBLEM 9-6

Predict the products using the electromotive or activity series for the following single-replacement reactions and balance:

(a) $Cd + Cu(NO_3)_2$ (aq) --->

(b) $Br_2 + NaI$ (aq) --->

(c) $Sn + NaCl$ (aq) --->

ANSWERS AND SOLUTIONS TO PROBLEM 9-6

(a) $Cd + Cu(NO_3)_2$ (aq) ---> $Cd(NO_3)_2 + Cu$

Cadmium is higher in the electromotive or activity series than copper and thus cadmium will replace copper from its salt. The cadmium salt that is formed is $Cd(NO_3)_2$ because cadmium when combined has a 2^+ ionic charge and nitrate is NO_3^-. The copper appears as a solid. The equation is balanced as written.

(b) $Br_2 + 2 NaI$ (aq) ---> $2 NaBr + I_2$

Bromine is higher in the halogen series than iodine and thus bromine will replace iodine from its salt (see periodic table). The bromine salt that is formed is NaBr because sodium has an ionic charge of 1^+ (Group IA,1) and bromine has an ionic charge of 1^- (Group VIIA, 7 - 8 = - 1 Iodine is a diatomic molecule and is written as I_2. The iodine is dissolved in the aqueous solution. Balance the equation and check each atom.

(c) $Sn + NaCl$ (aq) --->   NR

Tin is lower in the electromotive or activity series than sodium and thus there will be no reaction (NR).

If your answers are correct, proceed to Problem 9-7. If your answers are incorrect, review Section 9-8 (Examples 9-4, 9-5, 9-6, and 9-7 and Study Exercise 9-5) in your text.

Proceed to Problem 9-7.

PROBLEM 9-7

Predict the products using the rules for the solubility of inorganic substance in water for the following double-replacement reaction and balance them: [Indicate any precipitates by (s) and any gases by (g)].

(a) $Cu(NO_3)_2 + H_2S \longrightarrow$

(b) $FeCO_3 + HCl \longrightarrow$

(c) $Fe_2(SO_4)_3 + NaOH \longrightarrow$

ANSWERS AND SOLUTIONS TO PROBLEM 9-7

(a) $Cu(NO_3)_2 + H_2S \longrightarrow CuS(\underline{s}) + 2\ HNO_3$

The copper ion changes places with the hydrogen ion to form insoluble copper(II) sulfide (CuS, see rule 7), and the hydrogen ion reacts with the nitrate ion to form a new acid, nitric acid (HNO_3). Balance the equation by placing

a 2 in front of the HNO_3. Do not change the formula of

nitric acid to $H_2(NO_3)_2$ to balance the equation. Check each atom.

(b) $FeCO_3 + 2\ HCl \longrightarrow FeCl_2 + H_2O + CO_2(\underline{g})$

The iron(II) ion changes places with the hydrogen ion to form the salt, $FeCl_2$ which is soluble - rule 2. The

hydrogen ion reacts with the carbonate ion to form carbonic acid (H_2CO_3) which is unstable and decomposes to form

water and carbon dioxide. Balance the equation with a 2 in front of the HCl. Check each atom.

(c) $Fe_2(SO_4)_3 + 6\ NaOH \longrightarrow 2\ Fe(OH)_3(\underline{s}) + 3\ Na_2SO_4$

The iron(III) ion changes places with the sodium ion to form the base, $Fe(OH)_3$ which is insoluble - rule 6. The

sodium ion reacts with the sulfate ion to form Na_2SO_4.

Balance the equation by starting with Fe in $Fe_2(SO_4)_3$.

Place a 2 in front of $Fe(OH)_3$ and 3 in front of Na_2SO_4 to

balance the iron(III) and sulfate ions, respectively. Balance the sodium ion in the reactants by placing a 6 in front of NaOH which also balances the hydroxide ions. Check each atom.

If your answers are correct, proceed to Problem 9-8. If your answers are incorrect, review Section 9-9 (Examples 9-8 and 9-9 and Study Exercise 9-6) in your text.

Proceed to Problem 9-8.

PROBLEM 9-8

Predict the products for the following neutralization reactions and balance:

(a) $Fe(OH)_3$ + HCl --->

(b) HNO_3 + $Al(OH)_3$ --->

(c)   NaOH + SO_3 --->

ANSWERS AND SOLUTIONS TO PROBLEM 9-8

(a) $Fe(OH)_3$ + 3 HCl ---> $FeCl_3$ + 3 H_2O

The hydroxide ion reacts with the hydrogen ion to form water molecules, leaving the iron(III) ion and the chloride ion to form iron(III) chloride ($FeCl_3$) which is soluble - rule 2. Placing a 3 in front of HCl and a 3 in front of H_2O, gives a balanced equation. Check each atom.

(b)   3 HNO_3 + $Al(OH)_3$ ---> $Al(NO_3)_3$ + 3 H_2O

The hydrogen ion reacts with the hydroxide ion to form water molecules, leaving the aluminum ion and the nitrate ion to form aluminum nitrate [$Al(NO_3)_3$] which is soluble - rule 1. Placing a 3 in front of HNO_3 and a 3 in front of H_2O, gives a balanced equation. Check each atom.

(c)   2 NaOH + SO_3 ---> Na_2SO_4 + H_2O

The sulfur trioxide reacts with the sodium hydroxide to form water and a salt, Na_2SO_4. Sulfur trioxide forms the polyatomic ion, SO_4^{2-}; sulfur dioxide forms the polyatomic ion SO_3^{2-} (see Section 9-10, Study Hint) in your text.

From a knowledge of the ionic charge of sodium (Na^+) and

the ionic charge on the sulfate ion ($SO_4{}^{2-}$), you can write the correct formula of the salt: Na_2SO_4. Balance the equation by placing a 2 in front of $NaOH$. Check each atom.

If your answers are correct, proceed to Problem 9-9. If your answers are incorrect, review Section 9-10 (Examples 9-10 and 9-11 and Study Exercise 9-7) in your text.

Proceed to Problem 9-9.

PROBLEM 9-9

Balance the following reaction equations and classify each as a (i) combination reaction, (ii) decomposition reaction, (iii) single-replacement reaction, (iv) double-replacement reaction, or (v) neutralization reaction.

(a) $Ba(OH)_2(aq)$ + $H_2SO_4(aq)$ ----> $BaSO_4(s)$ + $H_2O(\ell)$

(b) $BaCO_3(s)$ ----> $BaO(s)$ + $CO_2(g)$

(c) $B(s)$ + $O_2(g)$ $\xrightarrow{\triangle}$ $B_2O_3(s)$

ANSWERS AND SOLUTIONS TO PROBLEM 9-9

(a) $Ba(OH)_2(aq)$ + $H_2SO_4(aq)$ ----> $BaSO_4(s)$ + 2 $H_2O(\ell)$

This is a reaction of a base with an acid and hence is a neutralization reaction (v). Notice that $BaSO_4$ is a precipitate, see rule 2, the exception.

(b) $BaCO_3(s)$ ----> $BaO(s)$ + $CO_2(g)$

$BaCO_3$ is a metal carbonate. Metal carbonates on heating decompose to produce carbon dioxide gas (see Section 9-7, number 1). This reaction is, therfore, a decomposition reaction (ii).

(c) 4 $B(s)$ + 3 $O_2(g)$ $\xrightarrow{\triangle}$ 2 $B_2O_3(s)$

Boron is a nonmetal (see periodic table). Nonmetals and oxygen react on heating to produce the nonmetal oxide (see Section 9-6, number 2). This reaction is, therefore, a combination reaction (i). In balancing this equation, place a 2 in front of the B as follows:

2 $B(s)$ + $O_2(g)$ $\xrightarrow{\triangle}$ $B_2O_3(s)$ (unbalanced)

To balance the oxygen atoms, place 3/2 in front of O_2, as follows:

$$2 \; B(\underline{s}) \; + \; 3/2 \; O_2(\underline{g}) \; \xrightarrow{\triangle} \; B_2O_3(\underline{s}) \quad \text{(balanced)}$$

The coeficients must be whole numbers, so multiply all coeficients by 2 and the correct balanced equation is as follows:

$$4 \; \overset{\checkmark}{B}(\underline{s}) \; + \; 3 \; \overset{\checkmark}{O_2}(\underline{g}) \; \xrightarrow{\triangle} \; 2 \; \overset{\checkmark\checkmark}{B_2O_3}(\underline{s}) \quad \text{(balanced)}$$

Check each atom.

If your answers are correct, you have completed this chapter. If your answers are incorrect, review Sections 9-5 through 9-10 in your text.

These problems conclude the chapter on chemical equations.

Now take the sample quiz to see if you have mastered the material in this chapter.

SAMPLE QUIZ

<u>Quiz #9</u>

You may use the periodic table, the electromotive or activity series, and the rules for the solubility of inorganic substances in water.

1. Balance the following equations:

(a) $Al(\underline{s}) + O_2(\underline{g}) \xrightarrow{\triangle} Al_2O_3(\underline{s})$

(b) $C_2H_6(\underline{g}) + O_2(\underline{g}) \xrightarrow{\triangle} CO_2(\underline{g}) + H_2O(\underline{\ell})$

2. Change the following equation into a chemical equation and and balance by inspection:

Calcium chloride + phosphoric acid --->

                    calcium phosphate(<u>s</u>) + hydrochloric acid

3. Classify each of the following balanced reaction equations as a (i) combination reaction, (ii) decomposition reaction (iii) single-replacement reaction, (iv) double-replacement reaction, or (v) neutralization reaction.

(a) $2 NaNO_3 (s) \xrightarrow{\triangle} 2 NaNO_2 (s) + O_2 (g)$

(b) $2 Al (s) + 6 HCl (aq) \longrightarrow 2 AlCl_3 (aq) + 3 H_2 (g)$

4. Complete and balance the following single-replacement, double-replacement, or neutralization reactions. [Indicate any precipitates by (s) and any gases by (g).]

(a) $Mg (s) + HCl (aq) \longrightarrow$

(b) $Pb(NO_3)_2 (aq) + HCl (aq) \xrightarrow{cold}$

Answers for Quiz #9

1. (a) 4 + 3 ---> 2; (b) 2 + 7 ---> 4 + 6

2. $3 CaCl_2 + 2 H_3PO_4 \longrightarrow Ca_3(PO_4)_2 (s) + 6 HCl$

3. (a) decomposition reaction (ii); (b) single-replacement reaction (iii)

4. (a) 1 + 2 ---> $MgCl_2 + H_2 (g)$

(b) 1 + 2 ---> $PbCl_2 (s) + 2 HNO_3$

Review Exam #3 (Chapters 7 to 9) [Answers are in ( ) to the right of each question.]

You may use the periodic table, the electromotive or activity series, and the rules for the solubility of inorganic substances in water.

Element	Atomic Mass Units (amu)
C	12.0
H	1.0
X	1.3
Y	3.0
Ca	40.1
O	16.0
Na	23.0
N	14.0

1. The correct formula for calcium chloride is:

    A. $CaCl$ (B)

    B. $CaCl_2$

    C. Ca_2Cl

    D. Ca_3Cl_2

    E. Ca_2Cl_3

2. The correct name for K_2CrO_4 is:

    A.   potassium chromite   (B)

    B.   potassium chromate

    C.   potassium dichromite

    D.   potassium dichromate

    E.   dipotassium chromate

3. The correct formula for tin(IV) sulfate is:

    A. $SnSO_4$ (C)

    B. $SnSO_3$

    C. $Sn(SO_4)_2$

    D. $Sn(SO_3)_2$

    E. $Sn_2(SO_4)_4$

4. The correct name of N_2O_4 is:

    A.   nitrogen oxide   (D)

    B.   nitrogen tetroxide

    C.   dinitrogen oxide

    D.   dinitrogen tetroxide

    E.   nitrogen dioxide

5. The correct formula for dichlorine heptoxide is:

    A. Cl_2O_7 (A)

    B. Cl_2O_6

    C. C_2O_7

    D. D_2O_6

    E. ClO_7

6. The common name for sodium hydrogen carbonate is:

    A. slaked lime   (D)

    B. lye

    C. vinegar

    D. baking soda

    E. chalk

7. The correct name for $NaMnO_4$ is:

    A. sodium manganate   (C)

    B. sodium sulfate

    C. sodium permanganate

    D. sodium manganite

    E. sodium permanganite

8. Which one of the following compounds is considered to be an acid? (Assume that all compounds are in water solution.)

    A. NaOH   (E)

    B. Na_3PO_4

    C. Na_2HPO_4

    D. $Al(OH)_3$

    E. H_3PO_4

9. Calculate the number of moles of methane (CH_4) in 8.00 g of methane.

   A.   5.00 mol   (D)

   B.   40.0 mol

   C.   0.615 mol

   D.   0.500 mol

   E.   8.00 mol

10. Calculate the number of molecules in 32.0 g of methane (CH_4).

    A. 1.20×10^{24} molecules   (A)

    B. 1.20×10^{23} molecules

    C. 3.01×10^{23} molecules

    D. 1.92×10^{25} molecules

    E. 1.92×10^{24} molecules

11. Calculate the number of grams of methane (CH_4) in 4.30 L of methane gas at STP.

    A.   1.62 g   (B)

    B.   3.07 g

    C.   0.192 g

    D.   0.269 g

    E.   2.50 g

12. Calculate the molar mass of a gas if 2.00 g of the gas occupies 1.12 L at STP.

    A.   4.00 g   (B)

    B.   40.0 g

    C.   10.0 g

    D. $1\overline{00}$ g

    E. $4\overline{00}$ g

138

13. Calculate the density in g/L of X_2Y gas at STP.

A.   0.25 g/L   (A)

B.   0.19 g/L

C.   4.0 g/L

D.   0.12 g/L

E.   1.9 g/L

14. Calculate the percent composition of calcium carbonate.

A.   40.1 % Ca; 36.0 % C; 48.0 % O   (D)

B.   40.1 % Ca; 12.0 % C; 16.0 % O

C.   40.1 % Ca; 12.0 % C; 33.3 % O

D.   40.1 % Ca; 12.0 % C; 48.0 % O

E.   58.7 % Ca; 17.6 % C; 23.5 % O

15. A certain compound gave the following chemical analysis:
27.1 % Na, 16.5 % N, and 56.5 % O. The empirical formula
for the compound is:

A.   NaNO   (E)

B. $NaNO_2$

C. Na_2NO_2

D. $Na_2N_2O_7$

E. $NaNO_3$

16. A certain compound gave the following chemical analysis:
85.7 % C and 14.3 % H. The empirical formula for the
compound is:

A.   CH   (B)

B. CH_2

C. C_2H

D. CH_3

E. C_2H_6

139

17. The molecular mass of the compound in question 16 is 56.0 amu. The molecular formula for the compound is:

    A. C_4H_4 (E)

    B. C_2H_4

    C. C_3H_6

    D. C_4H_{12}

    E. C_4H_8

18. Balance the following chemical equation by inspection.

    $$Al(\underline{s}) + O_2(\underline{g}) \xrightarrow{\triangle} Al_2O_3(\underline{s})$$

    The coefficients in this balanced equation are:

    A. 2 + 3 ---> 1   (C)

    B. 2 + 1 ---> 1

    C. 4 + 3 ---> 2

    D. 2 + 1 ---> 2

19. The equation in question 18 is an example of a:

    A. combination reaction   (A)

    B. decomposition reaction

    C. single replacement reaction

    D. double replacement reaction

    E. neutralization reaction

20. Change the following equation into a chemical equation and balance by inspection.

    phosphoric acid + potassium hydroxide --->

    potassium phosphate + water

    A. $HPO_4 + KOH ---> KPO_4 + H_2O$ (C)

    B. $H_3PO_4 + KOH ---> KPO_4 + 2\ H_2O$

    C. $H_3PO_4 + 3\ KOH ---> K_3PO_4 + 3\ H_2O$

    D. $H_3PO_4 + KOH ---> K_3PO_4 + H_2O$

140

21. Balance the following reaction equation. Classify it as a
    (i) combination reaction, (ii) decomposition reaction,
    (iii) single-replacement reaction, (iv) double-replacement
    reaction, or (v) neutralization reactions

$$KClO_3(s) \xrightarrow[\triangle]{MnO_2} KCl(s) + O_2(g)$$

    (B)

A. $KClO_3(s) \xrightarrow[\triangle]{MnO_2} KCl(s) + 3\ O_2(g)$, decomposition
                                                    reaction (ii)

B. $2\ KClO_3(s) \xrightarrow[\triangle]{MnO_2} 2\ KCl(s) + 3\ O_2(g)$, decomposition
                                                    reaction (ii)

C. $KClO_3(s) \xrightarrow[\triangle]{MnO_2} KCl(s) + 3\ O_2(g)$, combination
                                                    reaction (i)

D. $2\ KClO_3(s) \xrightarrow[\triangle]{MnO_2} 2\ KCl(s) + 3\ O_2(g)$, combination
                                                    reaction (i)

22. In the equation in question 21, the MnO_2 is a:

A.   product   (B)

B.   catalyst

C.   reactant

D.   form of heat

23. Complete and balance the following chemical equation:

$$Al(s) + CuCl_2(aq) \longrightarrow$$

A. $Al(s) + CuCl_2(aq) \longrightarrow AlCl_2(aq) + Cu(s)$ (B)

B. $2\ Al(s) + 3\ CuCl_2(aq) \longrightarrow 2\ AlCl_3(aq) + 3\ Cu(s)$

C. $Al(s) + CuCl_2(aq) \longrightarrow AlCl_3(aq) + Cu(s)$

D. $2\ Al(s) + 3\ CuCl_2(aq) \longrightarrow 2\ AlCl_3(aq) + Cu(s)$

24. Complete and balance the following chemical equation:

$$BaCl_2 + Na_2SO_4 \longrightarrow$$

A. $BaCl_2 + Na_2SO_4 \longrightarrow BaSO_4 + Na_2Cl_2$ (D)

B. $BaCl_2 + Na_2SO_4 \longrightarrow BaSO_4 + (NaCl)_2$

C. $BaCl_2 + Na_2SO_4 \longrightarrow BaSO_4 + NaCl$

D. $BaCl_2 + Na_2SO_4 \longrightarrow BaSO_4 + 2\ NaCl$

141

25. In the equation in question 24, the precipitate is:

    A. $BaCl_2$ (C)

    B. Na_2SO_4

    C. $BaSO_4$

    D.   NaCl

CHAPTER 10

CALCULATIONS INVOLVING CHEMICAL EQUATIONS. STOICHIOMETRY

Stoichiometry is measurement of the relative quantities of chemical reactants and products in a chemical reaction. In this chapter we will use the coefficients in the balanced equation (Chapter 9) to relate the amounts of reactants and products to each other. We will calculate the amounts of material or energy produced or required in a balanced chemical equation. In our calculations we will rely heavily on material you learned in Chapter 8, that is the calculations of molar masses, moles of units, and molar volume of a gas. We strongly urge you to review these topics.

SELECTED TOPICS

1. A balanced equation gives more information than simply which substances are reactants and which are products. It give the quantities of reactants and products in a given chemical reaction.

$$C_3H_8(g) + 5\ O_2(g) \xrightarrow{\triangle} 3\ CO_2(g) + 4\ H_2O(g)$$
propane

a. Reactants and products. C_3H_8 (propane) reacts with O_2 (oxygen) with sufficient heat ($\triangle$) to produce CO_2 (carbon dioxide) and H_2O (gaseous water).

b. Molecules of reactants and products. 1 molecule of C_3H_8 reacts with 5 molecules of O_2 to produce 3 molecules of CO_2 and 4 molecules of H_2O.

c. Moles of reactants and products. 1 mol of C_3H_8 reacts with 5 mol of O_2 to produce 3 mol of CO_2 and 4 mol of H_2O. This statement is just an expansion of the above statement with molecules; that is, 6.02×10^{23} molecules of C_3H_8 (1 mol) reacts with $5 \times 6.02 \times 10^{23}$ molecules of O_2 (5 mol) to produce $3 \times 6.02 \times 10^{23}$ molecules of CO_2 (3 mol) and $4 \times 6.02 \times 10^{23}$ of H_2O (4 mol).

d. Volumes of gases. 1 volume of C_3H_8 reacts with 5 volumes of O_2 to produce 3 volumes of CO_2 and 4 volumes of H_2O, if all volumes are gases and are measured at the same temperature and pressure.

e. Relative masses of reactants and products. 44.0 g of C_3H_8 (1 mol C_3H_8 x $\frac{44.0 \text{ g } C_3H_8}{1 \text{ mol } C_3H_8}$) reacts with 160 g of O_2

(5 mol O_2 x $\frac{32.0 \text{ g } O_2}{1 \text{ mol } O_2}$) to produce 132 g of CO_2

(3 mol CO_2 x $\frac{44.0 \text{ g } CO_2}{1 \text{ mol } CO_2}$) and 72.0 g of H_2O

(4 mol H_2O x $\frac{18.0 \text{ g } H_2O}{1 \text{ mol } H_2O}$). The sum of the reactants

(44.0 g + 160 g = 204 g) equals the sum of the products

(132 g + 72.0 g = 204 g), obeying the law of conservation of mass.

See Section 10-1 in your text.

2. In solving stoichiometry problems we will use the mole method, an extension of dimensional analysis we used in Chapters 2 and 8. The mole method involves three basic steps:

Step I: Moles of known. Calculate moles of units of elements, compounds, or ions from mass or volume (if gases) of the known substances.

Step II: Use coefficients - moles of unknown. Using the coefficients of the substances in the balanced equation, calculate moles of the unknown.

Step III: Mass or volume of unknown. From the moles of the unknown, calculate the mass or volume (if gases) of the unknown.

See Section 10-2 in your text.

3. There are three types of stoichiometry problems. They are:

(1) mass-mass
(2) mass-volume
(3) volume-volume

See Section 10-3 in your text.

4. In mass-mass stoichiometry problems, we give or ask for mass units of both the known and the unknown. In mass to mass examples we express the known in mass units and ask for the unknown in mass units. These examples involve all three basic steps. In mass to moles and vice versa examples we are again working with mass units, but sometimes the information is given in moles, so we do not need to calculate moles, or the unknown is asked for in moles, so we do not need to do a further calculation. Therefore, step I or III, may be eliminated. In moles to moles examples we are given moles of reactants or products and asked to calculate moles of product or reactant. Therefore, both steps I and III may be eliminated, but never step II. In limiting reagent examples, two amounts of reactants are given, but only one reactant acts as a limiting reagent. This limiting reagent is the first reactant to be entirely consumed. It limits the amount of product obtained. The other reactant is in excess and is called the excess reagent. The theoretical yield is the amount of product expected if all the limiting reagent forms product with none of it left. This assumes that none of the product is lost in isolation and purification. This is not the usual case. Some of the product is lost in side reactions, isolation, and purification. The amount of product obtained in an actual chemical reaction is called the actual yield. The percent yield is the percent of the theoretical yield that is actually obtained and is calculated as follows:

$$\% \text{ yield} = \frac{\text{actual yield}}{\text{theoretical yield}} \times 100$$

See Section 10-4 in your text.

5. In mass-volume stoichiometry problems, either the known or unknown is a gas. The gas will be measured at standard temperature and pressure [STP, $0^{\circ}C$ and 760 mm Hg (torr)]. This involves the molar volume of a gas - 22.4 L/1 mol of any gas at STP. See Section 10-5 in your text.

6. In volume-volume stoichiometry problems, the coefficients in the balanced equation represent volumes of gases. This assumes that substances in the balanced equation are gases and measured at the same temperature and pressure. The basis of this calculation is Gay-Lussac's law of combining volumes, which states that at the same temperature and pressure whenever gases react or gases are formed, they do so in the ratio of small whole numbers by volumes. This ratio of small whole numbers by volume is directly proportional to the values of the coefficients in the balanced equation. See Section 10-6 in your text.

7. In addition to the amounts of substances involved in chemical reactions, energy relationships are also important in chemical reactions. This energy is usually in the form of

heat energy. The heat of reaction is the number of calories or joules of heat energy given off (evolved) or absorbed in a particular chemical reaction. In exothermic reactions, heat energy is given off. If the reaction is carried out in a flask, the flask gets warm. In endothermic reactions, heat is absorbed. If the reaction is carried out in a flask, the flask gets cold. We can use the heat energy, either exothermic or endothermic, in stoichiometry problems. The quantity of heat energy is related to the moles of reactants or products in the balanced equation. We use the heat of reaction as we used moles in step II of our three basic steps. See Section 10-7 in your text.

PROBLEM 10-1

Calculate the number of grams of chlorine produced by the reaction of 42.1 g of potassium permanganate with excess hydrochloric acid. The balanced equation for the reaction is

$$2 \ KMnO_4(\underline{aq}) + 16 \ HCl(\underline{aq}) \longrightarrow$$

$$2 \ KCl(\underline{aq}) + 2 \ MnCl_2(\underline{aq}) + 5 \ Cl_2(\underline{g}) + 8 \ H_2O(\ell)$$

ANSWER AND SOLUTION TO PROBLEM 10-1

47.3 g   The molar mass of $KMnO_4$ is 158.0 and the molar mass of Cl_2 is 71.0 g. (Be sure to use the table of approximate atomic masses in the inside back cover of your text.) The known quantity is 42.1 g $KMnO_4$, and the unknown quantity is the grams of Cl_2 produced.

$$42.1 \ \cancel{g \ KMnO_4} \times \frac{1 \ \cancel{mol \ KMnO_4}}{158.0 \ \cancel{g \ KMnO_4}} \times \frac{5 \ \cancel{mol \ Cl_2}}{2 \ \cancel{mol \ KMnO_4}} \times \frac{71.0 \ g \ Cl_2}{1 \ \cancel{mol \ Cl_2}}$$

$$= 47.3 \ g \ Cl_2$$

146

If your answer is correct, proceed to Problem 10-2. If your answer is incorrect, review Section 10-4 (mass to mass examples, Examples 10-1 and 10-2 and Study Exercise 10-1) in your text.

Proceed to Problem 10-2.

PROBLEM 10-2

Calculate the number of moles of ammonia that would be formed by treating 423 g of ammonium sulfate with excess sodium hydroxide. The balanced equation for the reaction is

$$(NH_4)_2SO_4(aq) + 2\ NaOH(aq) \longrightarrow Na_2SO_4(aq) + 2\ H_2O(\ell) + 2\ NH_3(g)$$

ANSWER AND SOLUTION TO PROBLEM 10-2

6.40 mol   In this problem only steps I and II are necessary, because the problem asks for the number of moles of ammonia. The molar mass of ammonium sulfate is 132.1 g.

$$423\ g\ \cancel{(NH_4)_2SO_4} \times \frac{1\ mol\ \cancel{(NH_4)_2SO_4}}{132.1\ g\ \cancel{(NH_4)_2SO_4}} \times \frac{2\ mol\ NH_3}{1\ mol\ \cancel{(NH_4)_2SO_4}}$$

$$= 6.40\ mol\ NH_3$$

If your answer is correct, proceed to Problem 10-3. If your answer is incorrect, review Section 10-4 (mass to moles and vice versa, Examples 10-3 and 10-4 and Study Exercise 10-2) in your text.

Proceed to Problem 10-3.

PROBLEM 10-3

Consider the following balanced equation:

$$4 \text{ Al}(\underline{s}) + 3 \text{ O}_2(\underline{g}) \xrightarrow{\triangle} 2 \text{ Al}_2\text{O}_3(\underline{s})$$

If 0.350 mol of aluminum reacts with excess oxygen to form aluminum oxide solid, calculate the number of moles of aluminum oxide produced.

ANSWERS AND SOLUTIONS TO PROBLEM 10-3

0.175 mol   The known is given in moles and the unknown is asked for in moles, so only step II is needed; hence, we may eliminate both steps I and II.

$$0.350 \text{ mol Al} \times \frac{2 \text{ mol Al}_2\text{O}_3}{1 \text{ mol Al}} = 0.175 \text{ mol Al}_2\text{O}_3$$

If your answer is correct, proceed to Problem 10-4. If your answer is incorrect, review Section 10-4 (moles to moles examples, Example 10-5 and Study Exercise 10-3) in your text.

Proceed to Problem 10-4

PROBLEM 10-4

When a mixture of powdered manganese(II) oxide and powdered aluminum is ignited, the following reaction takes place:

$$3 \text{ MnO}(\underline{s}) + 2 \text{ Al}(\underline{s}) \longrightarrow 3 \text{ Mn}(\underline{s}) + \text{Al}_2\text{O}_3(\underline{s}) \text{ (Balanced equation)}$$

(a) Calculate the number of grams of manganese that could be obtained if 14.2 g of manganese(II) oxide is allowed to react with 3.31 g aluminum.

(b) Calculate the number of moles of excess reagent remaining at the end of the reaction.

(c) Calculate the percent yield if 8.75 g of manganese is actually obtained.

(a) _____

(b) _____

(c) _____

ANSWERS AND SOLUTIONS TO PROBLEM 10-4

(a) 10.1 g   The molar mass of MnO is 70.9 g and the molar
                     mass of Al is 27.0 g. We must first determine
                     which of the reactants, MnO or Al, is the
                     limiting reagent. Answer this question as
                     follows:

               1. Calculate the moles of each used as in
                   step I.

                   14.2 g̶ ̶M̶n̶O̶ x $\dfrac{1\ mol\ MnO}{70.9\ g̶\ ̶M̶n̶O̶}$ = 0.200 mol MnO

                   3.31 g̶ ̶A̶l̶ x $\dfrac{1\ mol\ Al}{27.0\ g̶\ ̶A̶l̶}$ = 0.123 mol Al

               2. Calculate the moles of product that could
                   be produced from each reactant as in step
                   II.

                   0.200 m̶o̶l̶ ̶M̶n̶O̶ x $\dfrac{3\ mol\ Mn}{3\ m̶o̶l̶\ ̶M̶n̶O̶}$ = 0.200 mol Mn

                   0.123 m̶o̶l̶ ̶A̶l̶ x $\dfrac{3\ mol\ Mn}{2\ m̶o̶l̶\ ̶A̶l̶}$ = 0.184 mol Mn

149

3. The reactant that gives the least number of moles of the product is the limiting reagent. Hence, in this problem, Al is the limiting reagent (0.200 mol vs 0.184 mol of Mn) and MnO is in excess. Using 0.184 mol Mn, calculated from the moles of Al, the number of grams of Mn that could be produced would be

$$0.184 \text{ mol Mn} \times \frac{54.9 \text{ g Mn}}{1 \text{ mol Mn}} = 10.1 \text{ g Mn}$$

(b) 0.016 mol   The amount of excess MnO is equal to 0.200 mol MnO present at the start of the reaction (see step I) minus the amount which is consumed by reacting with the limiting reagent (Al). The amount consumed is

$$0.123 \text{ mol Al} \times \frac{3 \text{ mol MnO}}{2 \text{ mol Al}} = 0.184 \text{ mol MnO}$$

and the amount in excess is

0.200 mol MnO present
- 0.184 mol MnO consumed
0.016 mol MnO in excess

(c) 86.6%   Calculate the percent yield from the theoretical in (a) and actual yield as follows:

$$\frac{8.75 \text{ g Mn}}{10.1 \text{ g Mn}} \times 100 = 86.6\%$$

If your answers are correct, proceed to Problem 10-6. If your answers are incorrect, review Section 10-4 (limiting reagent examples and percent yields, Examples 10-6, 10-7, 10-8, and 10-9, and Study Exercise 10-4) in your text.

Proceed to Problem 10-5.

PROBLEM 10-5

The following reaction occurs:

$$N_2(g) + 3 H_2(g) \xrightarrow{\triangle} 2 NH_3(g) \text{ (Balanced equation)}$$

(a) Calculate the number of grams of ammonia that could be obtained if 56.0 g of nitrogen is allowed to react with 15.0 g hydrogen.

(b) Calculate the number of moles of excess reagent remaining at the end of the reaction.

(c) Calculate the percent yield if 50.0 g of ammonia is actually obtained.

150

(a) _____

(b) _____

(c) _____

ANSWERS AND SOLUTIONS TO PROBLEM 10-5

(a) 68.0 g   The molar masses of N_2 and H_2 are 28.0 g and
                   2.0 g, respectively. First, we must determine
                   whether N_2 or H_2 is the limiting reagent.

         1. Calculate the moles of each used as in
             step I.

         56.0 g N_2 x $\dfrac{1 \text{ mol } N_2}{28.0 \text{ g } N_2}$ = 2.00 mol N_2

         15.0 g H_2 x $\dfrac{1 \text{ mol } H_2}{2.0 \text{ g } H_2}$ = 7.50 mol H_2

2. Calculate the moles of product that could be produced from each reactant as in step II.

$$2.00 \text{ mol } \cancel{N_2} \times \frac{2 \text{ mol NH}_3}{1 \text{ mol } \cancel{N_2}} = 4.00 \text{ mol NH}_3$$

$$7.50 \text{ mol } \cancel{H_2} \times \frac{2 \text{ mol NH}_3}{3 \text{ mol } \cancel{H_2}} = 5.00 \text{ mol NH}_3$$

3. The reactant that gives the least number of moles of product is the limiting reagent. Hence, in this problem N_2 is the limiting reagent (4.00 mol vs 5.00 mol of NH_3) and H_2 is in excess. Using 4.00 mol NH_3, calculated from the moles of N_2, the number of grams of NH_3 that could be produced would be

$$4.00 \text{ mol } \cancel{NH_3} \times \frac{17.0 \text{ g NH}_3}{1 \text{ mol } \cancel{NH_3}} = 68.0 \text{ g NH}_3$$

(b) 1.50 mol

The amount of excess H_2 is equal to 7.50 mol H_2 present at the start of the reaction (see step I) minus the amount which is consumed by reacting with the limiting reagent (N_2). The amount consumed is

$$2.00 \text{ mol } \cancel{N_2} \times \frac{3 \text{ mol H}_2}{1 \text{ mol } \cancel{N_2}} = 6.00 \text{ mol H}_2$$

and the amount in excess is

```
  7.50 mol H₂ present
- 6.00 mol H₂ consumed
  1.50 mol H₂ in excess
```

(c) 73.5%

Calculate the percent yield as follows:

$$\frac{50.0 \text{ g } \cancel{NH_3} = \text{actual yield}}{68.0 \text{ g } \cancel{NH_3} = \text{theoretical yield}} \times 100 = 73.5\%$$

If your answers are correct, proceed to Problem 10-6. If your answers are incorrect, re-read Section 10-4 (limiting reagent examples and percent yields) in your text. Re-work Problems 10-4 and 10-5.

Proceed to Problem 10-6.

PROBLEM 10-6

Calculate the number of liters of nitrogen oxide measured at STP that could be obtained from the reaction of 7.50 g of copper with excess nitric acid. The balanced equation for this reaction is

$$3 \text{ Cu(s)} + 8 \text{ HNO}_3\text{(aq) (dilute)} \longrightarrow$$

$$3 \text{ Cu(NO}_3)_2\text{(aq)} + 2 \text{ NO(g)} + 4 \text{ H}_2\text{O(g)}$$

ANSWER AND SOLUTION TO PROBLEM 10-6

1.76 L   The molar mass of Cu is 63.5 g. The conditions are STP; hence, in step III, you must use the relation 1 mol NO at STP occupies 22.4 L.

$$7.50 \text{ g Cu} \times \frac{1 \text{ mol Cu}}{63.5 \text{ g Cu}} \times \frac{2 \text{ mol NO}}{3 \text{ mol Cu}} \times \frac{22.4 \text{ L NO at STP}}{1 \text{ mol NO}}$$

$$= 1.76 \text{ L NO at STP}$$

If your answer is correct, proceed to Problem 10-7. If your answer is incorrect, review Section 10-5 (Examples 10-10, 10-11, 10-12, and 10-13 and Study Exercises 10-5, 10-6, and 10-7) in your text.

Proceed to Problem 10-7.

PROBLEM 10-7

Calculate the volume of ammonia gas in liters produced by the reaction of 7.25 liters of hydrogen gas with an excess of nitrogen gas. All volumes being measured at 300 atmospheres pressure and 500°C. The balanced equation is as follows:

$$3 \text{ H}_2\text{(g)} + \text{N}_2\text{(g)} \xrightarrow{\triangle} 2 \text{ NH}_3\text{(g)}$$

ANSWER AND SOLUTION TO PROBLEM 10-7

4.83 L   Because all volumes are measured at the same temperature and pressure, their volumes are related to their
coefficients in the balanced equation.

$$7.25 \; \cancel{L \; H_2} \times \frac{2 \; L \; NH_3}{3 \; \cancel{L \; H_2}} = 4.83 \; L \; NH_3$$

If your answer is correct, proceed to Problem 10-8. If your
answer is incorrect, review Section 10-6 (Examples 10-14 and
10-15 and Study Exercises 10-8 and 10-9) in your text.

Proceed to Problem 10-8.

PROBLEM 10-8

Given the following balanced equation:

$$C_6H_{12}O_6 \, (\underline{s}) + 6 \; O_2 \, (\underline{g}) \; \text{---}> 6 \; CO_2 \, (\underline{g}) + 6 \; H_2O \, (\underline{g}) + 673 \; kcal$$
glucose,
dextrose

(a)   Is the reaction exothermic or endothermic?

(b)   Calculate the number of kilocalories of heat energy produced in the reaction of 40.0 g of glucose with an excess
of oxygen gas. (This amount of glucose is approximately
equivalent to the carbohydrate in the average 12 ounce
soft drink.)

(a) _____

154

(b) _____

ANSWERS AND SOLUTIONS TO PROBLEM 10-8

(a)   exothermic   Heat is on the product side of the equation; therefore, heat is given off and the reaction is exothermic.

(b) $15\overline{0}$ kcal   The molar mass of $C_6H_{12}O_6$ is $18\overline{0}$ g. The relationship between glucose and the heat of reaction is 1 mol $C_6H_{12}O_6$ to 673 kcal. We, therefore, solve this problem using steps I and II.

$$40.0 \; \cancel{g \; C_6H_{12}O_6} \times \frac{1 \; \cancel{mol \; C_6H_{12}O_6}}{180 \; \cancel{g \; C_6H_{12}O_6}} \times \frac{673 \; kcal}{1 \; \cancel{mol \; C_6H_{12}O_6}}$$

$$= 15\overline{0} \; kcal$$

If your answers are correct, you have completed this chapter. If your answers are incorrect, review Section 10-7 (Problem Examples 10-16 and 10-17 and Study Exercise 10-10) in your text.

This problem concludes the chapter on stoichiometry.

Now take the sample quiz to see if you have mastered the material in this chapter.

SAMPLE QUIZ

Quiz #10

1. Calculate the number of moles of B required to produce 5.20 g of E, according to the following balanced equation:

A + 2 B ---> C + D + E
(molar mass E = 44.0 g)

2. Calculate the number of liters of E produced at STP if 0.625 mol of B reacts with A, according to the following balanced equation:

$$A + 2 B ---> C + D + E(g)$$

3. Calculate the number of kilocalories of heat energy produced by burning 37.0 g of methane (CH_4), according to the following balanced equation:

$$CH_4(g) + 2 O_2(g) ---> CO_2(g + 2 H_2O(g) + 213 \text{ kcal} \quad (\text{at } 25^{\circ}C)$$

(C = 12.0 amu, H = 1.0 amu)

4. Given the following balanced equation:

$$CH_4(g) + 2 O_2(g) \overset{\triangle}{==>} CO_2(g) + 2 H_2O(g)$$

(C = 12.0 amu, H = 1.0 amu, O = 16.0 amu)

(a) Calculate the number of grams of CO_2 that could be produced if 28.6 g of CH_4 reacts with 59.3 g of O_2.

(b)   Calculate the percent yield if 30.0 g of CO_2 is ac-
       tually obtained.

(c)   Calculate the number of moles of excess reagent re-
       maining at the end of the reaction.

Solutions and Answers for Quiz #10

1. 5.20 g E x $\dfrac{\text{1 mol E}}{\text{44.0 g E}}$ x $\dfrac{\text{2 mol B}}{\text{1 mol E}}$ = 0.236 mol B

2. 0.625 mol B x $\dfrac{\text{1 mol E}}{\text{2 mol B}}$ x $\dfrac{\text{22.4 L E (STP)}}{\text{1 mol E}}$ = 7.00 L E (STP)

3. 37.0 g CH_4 x $\dfrac{\text{1 mol } CH_4}{\text{16.0 g } CH_4}$ x $\dfrac{\text{213 kcal}}{\text{1 mol } CH_4}$ = 493 kcal

4. (a) 28.6 g CH_4 x $\dfrac{\text{1 mol } CH_4}{\text{16.0 g } CH_4}$ = 1.79 mol CH_4

        59.3 g O_2 x $\dfrac{\text{1 mol } O_2}{\text{32.0 g } O_2}$ = 1.85 mol O_2

        1.79 mol CH_4 x $\dfrac{\text{1 mol } CO_2}{\text{1 mol } CH_4}$ = 1.79 mol CO_2

        1.85 mol O_2 x $\dfrac{\text{1 mol } CO_2}{\text{2 mol } O_2}$ = 0.925 mol CO_2 <--- Use

        0.925 mol CO_2 x $\dfrac{\text{44.0 g } CO_2}{\text{1 mol } CO_2}$ = 40.7 g CO_2

    (b) $\dfrac{\text{30.0 g } CO_2}{\text{40.7 g } CO_2}$ x 100 = 73.7%

    (c) Moles of CH_4 consumed.

        1.85 mol O_2 x $\dfrac{\text{1 mol } CH_4}{\text{2 mol } O_2}$ = 0.925 mol CH_4 consumed

           1.79   mol CH_4 present
        $-$ 0.925 mol CH_4 consumed
          _____
           0.865 mol CH_4 excess

        0.86 mol CH_4 excess to hundredths place.

157

CHAPTER 11

GASES

There are three physical states of matter. These states are: solid, liquid, and gas. In this chapter we will consider the gaseous state of matter, and our discussion will include the various gas laws. These laws relate the effects of the variables of a gas - volume, pressure, temperature, and mass - to each other.

SELECTED TOPICS

1. Pressure is force per unit area. Standard pressure is the pressure of the atmospheric gases at sea level. Standard pressure is equivalent to the pressure exerted by a column of mercury 76.0 cm in height. It is also equivalent to 760 mm mercury, 760 torr, or 1.00 atmosphere. There are other equivalent standard pressure units, but we will use centimeters of mercury, millimeters of mercury, torr, and atmospheres in this book. You must know the standard pressure in these units. See Section 11-2 in your text.

2. Boyle's law states that at constant temperature the volume of a fixed mass of a given gas is inversely proportional to its pressure. Mathematically Boyle's law is expressed as

$$V \propto \frac{1}{P} \quad \text{(temperature constant)}$$

Introducing a constant of proportionality (k), we can write an equation as follows:

$$V = k \times \frac{1}{P}, \text{ or}$$

$$PV = k \tag{11-1}$$

Because P x V is equal to a constant (k), we can express different conditions of pressure and volume for the same mass of gas at constant temperature:

$$P_{new} \times V_{new} = k = P_{old} \times V_{old} \tag{11-2}$$

From this equation, we can solve for a new pressure or a new volume as follows:

$$P_{new} = P_{old} \times V_{factor} \tag{11-3}$$

$$V_{new} = V_{old} \times P_{factor} \tag{11-4}$$

We can evaluate the V_{factor} and P_{factor} by considering the effect the change in volume or pressure has on the old pressure or volume, and how this change will effect the new pressure or volume. When solving problems involving Boyle's law, you reason that increasing the pressure will decrease the volume and decreasing the pressure will increase the volume. Use similar reasoning to consider the effect of changing volume on pressure. Using this reasoning process means that you do not have to memorize a formula. See Section 11-3 in your text.

3. Charles's law states that at constant pressure the volume of a fixed mass of a given gas is directly proportional to the temperature in kelvins. We express Charles' law mathematically as:

$$V \propto T \text{ (pressure constant)}$$

$$V = \underline{k}T \ (\underline{k} = \text{proportionality constant)}, \text{ or}$$

$$\frac{V}{T} = \underline{k} \qquad\qquad (11\text{-}5)$$

$$\frac{V_{new}}{T_{new}} = \underline{k} = \frac{V_{old}}{T_{old}} \qquad\qquad (11\text{-}6)$$

$$T_{new} = T_{old} \times V_{factor} \qquad\qquad (11\text{-}7)$$

$$V_{new} = V_{old} \times T_{factor} \qquad\qquad (11\text{-}8)$$

Important -- we must express the temperature in kelvins. Again use the reasoning process in working these problems rather than memorizing an equation. See Section 11-4 in your text.

4. Gay-Lussac's law states that at constant volume the pressure of a fixed mass of a given gas is directly proportional to the temperature in kelvins. We express Gay-Lussac's law mathematically as:

$$P \propto T \text{ (volume constant)}$$

$$P = \mathbf{k}T \ (\mathbf{k} = \text{proportionality constant)}, \text{ or}$$

$$\frac{P}{T} = \mathbf{k} \qquad\qquad (11\text{-}9)$$

$$\frac{P_{new}}{T_{new}} = \mathbf{k} = \frac{P_{old}}{T_{old}} \qquad\qquad (11\text{-}10)$$

$$P_{new} = P_{old} \times T_{factor} \qquad\qquad (11\text{-}11)$$

$$T_{new} = T_{old} \times P_{factor} \qquad\qquad (11\text{-}12)$$

See Section 11-5 in your text.

5. The combined gas law is obtained by combining Boyle's and Charles' laws into one mathematical expression:

$$\frac{P_{new}V_{new}}{T_{new}} = \frac{P_{old}V_{old}}{T_{old}} \quad \text{(fixed mass)} \quad (11-13)$$

Solving this equation for V_{new}, P_{new}, and T_{new} gives

$$V_{new} = V_{old} \times P_{factor} \times T_{factor} \quad (11-14)$$

$$P_{new} = P_{old} \times V_{factor} \times T_{factor} \quad (11-15)$$

$$T_{new} = T_{old} \times V_{factor} \times P_{factor} \quad (11-16)$$

We must consider separately each factor in these equations and its effect on the new volume, pressure, or temperature. The result depends on the magnitude of the separate factors. See Section 11-6 in your text.

6. Dalton's law of partial pressures states that each gas in a mixture of gases exerts a partial pressure equal to the pressure it would exert if it were the only gas present in the same volume. The total pressure is the sum of the partial pressures of all the gases present. We express this law mathematically as:

$$P_{total} = P_1 + P_2 + P_3 + \ldots P_x \quad (11-17)$$

P_1, P_2, P_3, P_x are the partial pressures of the individual gases in the mixture. The collection of a gas over water is an application of Dalton's law of partial pressures. The gas will contain a certain amount of water vapor. The total pressure of this wet gas is equal to the gas pressure plus the vapor pressure of water at the temperature at which the gas is collected. This equation is as follows:

$$P_{total} = P_{gas} + P_{water} \quad (11-18)$$

Calculate the pressure of the dry gas by subtracting from the total pressure, the known vapor pressure of water at the temperature at which the gas is collected.

$$P_{gas} = P_{total} - P_{water} \quad (11-19)$$

The vapor pressure of water at various temperatures is in Appendix V of your text. In all gas problems in which the gas is collected over water, subtract the vapor pressure of water at its given temperature from the total pressure to obtain the pressure of the dry gas. See Section 11-7 in your text.

7. In the ideal-gas equation, we can vary temperature, pressure, volume, and also mass. The ideal-gas equation is as follows:

$$PV = nRT \qquad (11-20)$$

P = pressure, V = volume, n = quantity of gas in moles, T = temperature, and R = the universal gas constant. The value of R that we shall use is 0.0821 atm • L/mol • K. You must know the ideal-gas equation and the numerical value of R and its units to work problems involving this equation. In using the value of R equal to 0.0821 atm • L/mol • K, you must use the pressure in atmospheres, the volume in liters, the quantity of gas in moles, and the temperature in kelvins. If these values are not given in these units, you must convert the values given to these units before you can use the ideal-gas equation with R equal to 0.0821 atm L/mol K. See Section 11-8 in your text.

8. We can solve various types of problems related to gases by using the gas laws. These types of problems are (a) molecular mass and molar mass problems, (b) density problems, and (c) stoichiometry problems.

In molecular mass and molar mass problems of a gas in Chapter 8, we solved for grams per mole which is numerically equal to the molecular mass in amu and the molar mass in grams. This calculation involved the use of a given mass of gas, its volume at STP, and the molar volume, 22.4 L/mol of any gas at STP (0°C and 760 torr). Now, using the gas laws, we can measure the volume of the gas at non-STP conditions and then convert this volume to STP conditions. We then complete the calculations as we did in Chapter 8, using the calculated STP volume of the gas and the molar volume.

In density problems of a gas in Chapter 8, we solved for grams per liter of the gas at STP conditions using the molar volume. Now, we can calculate the density of a gas at non-STP conditions using the gas laws. We calculate the density of the gas at STP conditions as we did previously (see Section 8-3 in your text), but now we convert the volume of 1.00 L at STP to a volume at non-STP conditions using the gas laws. Using this new volume, and the mass in grams from the density of the gas at STP, we can calculate the density of the gas at non-STP conditions. Only the volume and not the mass of a gas changes with changes in temperature and pressure.

In stoichiometry problems in Chapter 10, we expressed the volume of a gas at STP conditions. By using the gas laws, we can convert this volume at STP to non-STP conditions. See Section 11-9 in your text.

PROBLEM 11-1

A _sample of gas occupies a volume of 115 mL at a pressure of 720 mm Hg and a temperature of $0^{\circ}C$. Calculate its volume in mL at STP.

ANSWER AND SOLUTION TO PROBLEM 11-1

109 mL   STP conditions are $76\overline{0}$ mm Hg and $0^{\circ}C$.   The temperature is constant.

$$V_{old} = 115 \text{ mL} \qquad P_{old} = 72\overline{0} \text{ mm Hg} \quad \vdots \quad \text{pressure increases;}$$

$$V_{new} = ? \qquad P_{new} = 76\overline{0} \text{ mm Hg} \downarrow \quad \text{volume decreases}$$

$$V_{new} = V_{old} \times P_{factor}$$

The pressure has increased from $72\overline{0}$ mm Hg to $76\overline{0}$ mm Hg; hence, the new volume will be decreased, and we must write the pressure factor so that the new volume will show a decrease.   To show this decrease, we must write the pressure factor so that the ratio of pressures is less than 1 -- hence $\dfrac{72\overline{0} \text{ mm Hg}}{76\overline{0} \text{ mm Hg}}$.

$$V_{new} = 115 \text{ mL} \times \frac{72\overline{0} \text{ mm Hg}}{76\overline{0} \text{ mm Hg}} = 109 \text{ mL}$$

If your answer is correct, proceed to Problem 11-2. If your answer is incorrect, review Section 11-3 (Examples 11-1 and 11-2 and Study Exercise 11-1) in your text.

Proceed to Problem 11-2.

PROBLEM 11-2

A gas occupies a volume of 675 mL at $3\overline{0}^{\circ}C$. At what temperature in $^{\circ}C$ will the volume be 455 mL, the pressure remaining constant?

ANSWER AND SOLUTION TO PROBLEM 11-2

-69°C   P = constant

$$V_{old} = 675 \text{ mL}$$ volume decreases; $t_{old} = 3\overline{0}^\circ$C $T_{old} = 3\overline{0} + 273$
$$= 303 \text{ K}$$

$$V_{new} = 455 \text{ mL}$$ temperature decreases; $t_{new} = ?$ $T_{new} = ?$

$$T_{new} = T_{old} \times V_{factor}$$

The volume decreases; therefore, the new temperature decreases, and we must write the volume factor so that the new temperature will be less. We must write the ratio of volumes so that the ratio is less than 1 -- hence, $\dfrac{455 \text{ mL}}{675 \text{ mL}}$.

$$T_{new} = 303 \text{ K} \times \frac{455 \text{ mL}}{675 \text{ mL}} = 204 \text{ K}$$

Then convert this temperature in kelvins to $^\circ$C by subtracting the constant, 273.

$$204 \text{ K} = (204 - 273)^\circ\text{C} = -69^\circ\text{C}$$

Be sure to convert the temperature in kelvins to $^\circ$C because those are the units requested in the problem.

If your answer is correct, proceed to Problem 11-3. If your answer is incorrect, review Section 11-4 (Examples 11-3 and 11-4 and Study Exercise 11-2) in your text.

Proceed to Problem 11-3.

PROBLEM 11-3

A 5.00 L sample of a gas at $25\overline{0}^\circ$C and $7\overline{00}$ mm Hg is cooled to exactly 0°C at constant volume. Calculate its pressure in mm Hg.

ANSWER AND SOLUTION TO PROBLEM 11-3

365 mm Hg   V = constant

P_{old} = $\overline{700}$ torr t_{old} = $25\overline{0}^{o}C$

P_{new} = ? t_{new} = $0^{o}C$

T_{old} = $25\overline{0}$ + 273 = 523 K   ⋮ temperature decreases;

T_{new} =   0 + 273 = 273 K   ↓ pressure decreases

P_{new} = P_{old} x T_{factor}

Because the temperature decreases, the pressure decreases, and we must write the temperature factor
so that the new pressure will be less. To show this
decrease, we must write the temperature factor so
that the ratio of temperature is less than 1 - hence
$\frac{273 \text{ K}}{523 \text{ K}}$.

P_{new} = $\overline{700}$ mm Hg x $\frac{273 \text{ K}}{523 \text{ K}}$ = 365 mm Hg

If your answer is correct, proceed to Problem 11-4. If your
answer is incorrect, review Section 11-5 (Example 11-5 and
Study Exercise 11-3) in your text.

Proceed to Problem 11-4.

PROBLEM 11-4

The volume of a sample of gas is $\overline{500}$ mL at $6\overline{40}$ torr and $30^{o}C$.
Calculate the volume of the gas in mL at STP.

164

ANSWER AND SOLUTION TO PROBLEM 11-4

379 mL   STP conditions are $76\overline{0}$ torr and $0°C$

$$V_{old} = 5\overline{00} \text{ mL} \quad P_{old} = 64\overline{0} \text{ torr} \mid \text{ pressure increases;}$$

$$V_{new} = \quad ? \qquad P_{new} = 76\overline{0} \text{ torr} \downarrow \text{ volume decreases}$$

$$T_{old} = 3\overline{0} + 273 = 303 \text{ K} \mid \text{ temperature decreases;}$$

$$T_{new} = \quad 0 + 273 = 273 \text{ K} \downarrow \text{ volume decreases}$$

$$V_{new} = V_{old} \times P_{factor} \times T_{factor}$$

Because the pressure increases, this volume decreases, and the ratio of pressures will be less than 1. A decrease in temperature will also result in a volume decrease and the ratio of temperatures will also be less than 1.

$$V_{new} = 5\overline{00} \text{ mL} \times \frac{64\overline{0} \text{ torr}}{760 \text{ torr}} \times \frac{273 \text{ K}}{303 \text{ K}} = 379 \text{ mL}$$

If your answer is correct, proceed to Problem 11-5. If your answer is incorrect, review Section 11-6 (Example 11-6 and 11-7 and Study Exercise 11-4) in your text.

Proceed to Problem 11-5.

PROBLEM 11-5

The volume of a certain gas, collected over water, is 174 mL at $27°C$ and 640.0 torr. Calculate the volume in mL of the dry gas at STP. (See Appendix V in your text for the vapor pressure of water.)

_____

ANSWER AND SOLUTION TO PROBLEM 11-5

128 mL   STP conditions are $76\overline{0}$ torr and $0°C$. First, determine the pressure of the dry gas at the initial volume of 174 mL and temperature of $27°C$. From Appendix V in your text, the vapor pressure of water at $27°C$ is 26.7 torr. Calculate the pressure of the dry gas as follows:

$$P_{\text{dry gas}} = P_{\text{total}} - P_{\text{water}} = 640.0 \text{ torr} - 26.7 \text{ torr}$$
$$= 613.3 \text{ torr}$$

If the water vapor were removed - that is, if the gas were dry - the pressure of the gas would measure 613.3 torr in a volume of 174 mL at 27°C. Next, work a combined gas law problem as follows:

$V_{\text{old}} = 174 \text{ mL}$ $P_{\text{old}} = 613.3 \text{ torr}$ | pressure increases;

$V_{\text{new}} = $? $P_{\text{new}} = 76\overline{0}$ torr ↓ volume decreases

$T_{\text{old}} = 27 + 273 = 30\overline{0} \text{ K}$ ¦ temperature decreases;

$T_{\text{new}} = 0 + 273 = 273 \text{ K}$ ↓ volume decreases

$V_{\text{new}} = V_{\text{old}} \times P_{\text{factor}} \times T_{\text{factor}}$

$V_{\text{new}} = 174 \text{ mL} \times \dfrac{613.3 \text{ torr}}{760 \text{ torr}} \times \dfrac{273 \text{ K}}{30\overline{0} \text{ K}} = 128 \text{ mL}$

If your answer is correct, proceed to Problem 11-6. If your answer is incorrect, review Section 11-7 (Examples 11-8 and 11-9 and Study Exercise 11-5) in your text.

Proceed to Problem 11-6.

PROBLEM 11-6

Calculate the volume in liters of 0.345 mol of oxygen gas at $3\overline{0}$ °C and $63\overline{0}$ torr.

ANSWER AND SOLUTION TO PROBLEM 11-6

10.4 L   Using the ideal-gas equation, PV = nRT, and solving for V (volume), we obtain the following equation:

$V = \dfrac{nRT}{P}$

We must express the pressure in atmospheres, because in the value of $R = 0.821 \dfrac{atm \cdot L}{mol \cdot K}$, atmospheres is used.

$$630 \text{ torr} \times \frac{1 \text{ atm}}{760 \text{ torr}} = 0.829 \text{ atm}$$

Substituting the value for n (0.345 mol), R (0.821 atm L/mol K), T ($30°C + 273 K = 303 K$), and P (0.829 atm), we obtain the following:

$$V = \frac{0.345 \text{ mol} \times 0.0821 \dfrac{atm \cdot L}{mol \cdot K} \times 303 \text{ K}}{0.829 \text{ atm}} = 10.4 \text{ L}$$

If your answer is correct, proceed to Problem 11-7. If your answer is incorrect, review Section 11-8 (Examples 11-10, 11-11, and 11-12 and Study Exercise 11-6) in your text.

Proceed to Problem 11-7.

PROBLEM 11-7

Calculate the molecular mass and molar mass of dimethyl ether if 1.32 L of the gas at $25.0°C$ and $64\overline{0}$ torr has a mass of 2.09 g.

ANSWER AND SOLUTION TO PROBLEM 11-7

45.9 amu, (1) We must correct the volume of 1.32 L at $25.0°C$
45.9 g and $64\overline{0}$ torr to STP conditions ($0°C$ and $76\overline{0}$ torr, so that we can use the molar volume of a gas in the next step.

V_{old} = 1.32 L T_{old} = 25.0 + 273 = 298 K ┊ temperature
                                                      ┊ decreases;
V_{new} =   ? T_{new} =   0 + 273 = 273 K ┊ volume
                                                      ┊ decreases

P_{old} = 64$\overline{0}$ torr ┊ pressure increases;

P_{new} = 76$\overline{0}$ torr ↓ volume decreases

V_{new} = V_{old} x P_{factor} x T_{factor}

V_{new} = 1.32 L x $\dfrac{273 \text{ K̶}}{298 \text{ K̶}}$ x $\dfrac{64\overline{0} \text{ t̶o̶r̶r̶}}{76\overline{0} \text{ t̶o̶r̶r̶}}$ = 1.02 L (at STP)

(2) Calculate the molecular mass and molar mass, solving for the units g/mol:

$\dfrac{2.09 \text{ g}}{1.02 \text{ L̶(̶S̶T̶P̶)̶}}$ x $\dfrac{22.4 \text{ L̶(̶S̶T̶P̶)̶}}{1 \text{ mol}}$ = 45.9 g/mol,

molecular mass = 45.9 amu

molar mass = 45.6 g

If your answer is correct, proceed to Problem 11-8. If your answer is incorrect, review Section 11-9 (molecular mass and molar mass problems, Examples 11-13 and 11-14 and Study Exercise 11-7) in your text:

Proceed to Problem 11-8.

PROBLEM 11-8

Calculate the density of nitrogen gas in grams per liter at 72$\overline{0}$ torr and 27°C.

ANSWER AND SOLUTION TO PROBLEM 11-8

1.08 g/L   (1) Solve for the density of the gas at STP. The formula of nitrogen gas is N_2 and the molar mass is 28.0 g. Calculate the density of the gas at STP as:

$$\frac{28.0 \text{ g } N_2}{1 \text{ mol } N_2} \times \frac{1 \text{ mol } N_2}{22.4 \text{ L } N_2 \text{ (STP)}} = 1.25 \text{ g/L (at STP)}$$

(2) Correct 1.00 L at STP to $72\overline{0}$ torr and $27^{O}C$.

$V_{old} = 1.00$ L $P_{old} = 76\overline{0}$ torr ⎰ pressure decreases;

$V_{new} = ?$ $P_{new} = 72\overline{0}$ torr ↓ volume increases

$T_{old} = 0 + 273 = 273$ K ⎰ temperature increases;

$T_{new} = 27 + 273 = 3\overline{00}$ K ↓ volume increases

$V_{new} = V_{old} \times P_{factor} \times T_{factor}$

$V_{new} = 1.00 \text{ L} \times \dfrac{76\overline{0} \text{ torr}}{720 \text{ torr}} \times \dfrac{3\overline{00} \text{ K}}{273 \text{ K}} = 1.16$ L

(3) Calculate the density at $72\overline{0}$ torr and $27^{O}C$, knowing that the 1.25 g of the gas which occupied 1.00 L at STP would occupy a volume of 1.16 L at $72\overline{0}$ torr and $27^{O}C$.

$$\frac{1.25 \text{ g}}{1.16 \text{ L}} = 1.08 \text{ g/L}$$

If our answer is correct, proceed to Problem 11-9. If your answer is incorrect, review Section 11-9 (density problems, Example 11-15 and Study Exercise 11-8) in your text.

Proceed to Problem 11-9.

PROBLEM 11-9

Calculate the volume of oxygen in liters measured at $27^{O}C$ and $64\overline{0}$ torr which could be obtained by heating 8.55 g of sodium nitrate according to the following balanced equation:

$$2 \text{ NaNO}_3 \xrightarrow{\triangle} 2 \text{ NaNO}_2 + O_2 (g)$$

1.47 L   (1) The equation is given and is balanced. Calculate the volume in liters of oxygen at STP. The molar mass of $NaNO_3$ is 85.0 g.

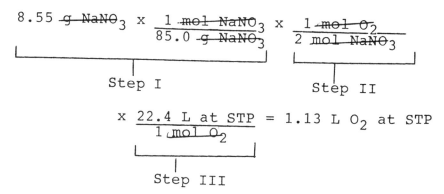

$$8.55 \text{ g } NaNO_3 \times \frac{1 \text{ mol } NaNO_3}{85.0 \text{ g } NaNO_3} \times \frac{1 \text{ mol } O_2}{2 \text{ mol } NaNO_3}$$

Step I   Step II

$$\times \frac{22.4 \text{ L at STP}}{1 \text{ mol } O_2} = 1.13 \text{ L } O_2 \text{ at STP}$$

Step III

(2) Correcting 1.13 L O_2 at STP to 27°C and 64$\overline{0}$ torr, we have

V_{old} = 1.13 L P_{old} = 76$\overline{0}$ torr ┊ pressure decreases;

V_{new} = ? P_{new} = 64$\overline{0}$ torr ↓ volume increases

T_{old} = 0 + 273 = 273 K ┊ temperature increases;

T_{new} = 27 + 273 = 30$\overline{0}$ K ↓ volume increases

V_{new} = V_{old} × P_{factor} × T_{factor}

$$V_{new} = 1.13 \text{ L} \times \frac{76\overline{0} \text{ torr}}{64\overline{0} \text{ torr}} \times \frac{30\overline{0} \text{ K}}{273 \text{ K}}$$

$$= 1.47 \text{ L } O_2 \text{ at } 27°C \text{ and } 64\overline{0} \text{ torr}$$

If your answer is correct, you have completed this chapter. If your answer is incorrect, review Section 11-9, stoichiometry problems, (Examples 11-16 and 11-17 and Study Exercise 11-9) in your text.

Proceed to Problem 11-10

PROBLEM 11-10

Calculate the number of grams of sodium nitrate needed to pro-duce 65$\overline{0}$ mL of oxygen at 27° C and 64$\overline{0}$ torr according to the following balanced equation:

2 NaNO$_3$ $\xrightarrow{\triangle}$ 2 NaNO$_2$ + O$_2$(g)

ANSWER AND SOLUTION TO PROBLEM 11-10

3.78 g (1) The equation is given and is balanced. Calculate the volume of oxygen in mL at STP conditions. The reason we calculate this volume at STP is so we can use the molar volume (22.4 L any gas at STP/1 mol) in step (2).

V_{old} = $65\overline{0}$ mL P_{old} = $64\overline{0}$ torr   pressure increases

V_{new} = ? P_{new} = $76\overline{0}$ torr ↓   volume decreases

T_{old} = 27 + 273 = $3\overline{00}$ K   temperature decreases

T_{new} = 0 + 273 = 273 K ↓   volume decreases

$$V_{new} = 65\overline{0} \text{ mL} \times \frac{64\overline{0} \text{ torr}}{76\overline{0} \text{ torr}} \times \frac{273 \text{ K}}{3\overline{00} \text{ K}} = 498 \text{ mL } O_2 \text{ at STP}$$

(2) Calculate the mass in grams of $NaNO_3$ from the volume of oxygen at STP. The molar mass of $NaNO_3$ is 85.0 g.

$$498 \text{ mL } O_2 \text{ (STP)} \times \frac{1 \text{ L}}{1000 \text{ mL}} \times \frac{1 \text{ mol } O_2}{22.4 \text{ L } O_2 \text{ (STP)}}$$

$$\times \frac{2 \text{ mol } NaNO_3}{1 \text{ mol } O_2} \times \frac{85.0 \text{ g } NaNO_3}{1 \text{ mol } NaNO_3} = 3.78 \text{ g } NaNO_3$$

If your answer is correct, you have completed this chapter. If your answer is incorrect, re-read Section 11-9 (stoichiometry problems) in your text. Re-work Problems 11-9 and 11-10.

These problems conclude the chapter on gases.

Now take the sample quiz to see if you have mastered the material in this chapter.

SAMPLE QUIZ

Quiz #11

Element	Atomic mass (amu)
K	39.1
Cl	35.5
O	16.0

1. A certain gas occupies 546 mL at 1.00 atm and $0^{\circ}C$. What volume in milliliters will it occupy at 10.0 atm and $100^{\circ}C$?

2. A gas occupies a volume of 6.00 L at $27^{\circ}C$. At what temperature in degrees Celsius ($^{\circ}C$) would the volume be 4.00 L, the pressure remaining constant?

3. Exactly 760 mL of oxygen is collected over water at $27^{\circ}C$ and 627.0 torr. Calculate the volume of dry oxygen at STP. The vapor pressure of water at $27^{\circ}C$ is 26.7 torr.

4. Calculate the molecular mass and molar mass of a gas if 1.25 g has a volume of 550 mL at $27.0^{\circ}C$ and 700 torr.

5. Calculate the volume of oxygen in liters produced at $64\overline{0}$ torr and $27^{\circ}C$ if 6.13 g potassium chlorate is heated according to the following <u>unbalanced</u> equation:

$$KClO_3(\underline{s}) \xrightarrow{\triangle} KCl(\underline{s}) + O_2(\underline{g}) \quad \text{UNBALANCED}$$

<u>Solutions</u> <u>and</u> <u>Answers</u> <u>for</u> <u>Quiz</u> #11

1. $V_{new} = 546 \text{ mL} \times \dfrac{1.00 \; \text{atm}}{10.0 \; \text{atm}} \times \dfrac{373 \; K}{273 \; K} = 74.6 \text{ mL}$

2. $T_{new} = 300 \text{ K} \times \dfrac{4.00 \; L}{6.00 \; L} = 200 \text{ K}$

   $200 \text{ K} - 273 = -73^{\circ}C$

3. $P_{dry} = 627.0 \text{ torr} - 26.7 \text{ torr} = 600.3 \text{ torr}$

   $V_{new} = 76\overline{0} \text{ mL} \times \dfrac{273 \; K}{300 \; K} \times \dfrac{600.3 \; \text{torr}}{760 \; \text{torr}} = 546 \text{ mL}$

4. $V_{new} = 55\overline{0} \text{ mL} \times \dfrac{273 \; K}{300 \; K} \times \dfrac{70\overline{0} \; \text{torr}}{760 \; \text{torr}} = 461 \text{ mL at STP}$

   $\dfrac{1.25 \text{ g}}{461 \text{ mL (STP)}} \times \dfrac{1000 \text{ mL}}{1 \text{ L}} \times \dfrac{22.4 \text{ L (STP)}}{1 \text{ mol}} = 60.7 \text{ g/mol},$
   $60.7 \text{ amu}, \; 60.7 \text{ g}$

5. $2 \; KClO_3(\underline{s}) \xrightarrow{\triangle} 2 \; KCl(\underline{s}) + 3 \; O_2(\underline{g})$

   $6.13 \; \text{g } KClO_3 \times \dfrac{1 \text{ mol } KClO_3}{122.6 \text{ g } KClO_3} \times \dfrac{3 \text{ mol } O_2}{2 \text{ mol } KClO_3} \times \dfrac{22.4 \text{ L (STP)}}{1 \text{ mol } O_2}$

   $= 1.68 \text{ L (STP)}$

   $V_{new} = 1.68 \text{ L} \times \dfrac{30\overline{0} \; K}{273 \; K} \times \dfrac{76\overline{0} \; \text{torr}}{64\overline{0} \; \text{torr}}$

   $= 2.19 \text{ L at } 64\overline{0} \text{ torr and } 27^{\circ}C$

CHAPTER 12

LIQUIDS AND SOLIDS

In this chapter we continue our discussion of the three physical states of matter, considering now liquids and solids. We
will be primarily concerned with energy measurements in changing physical states.

SELECTED TOPICS

1. <u>Condensation</u> is the conversion of vapor (gas) molecules in-
    to a liquid. <u>Evaporation</u> is the escape of molecules from
    the surface of the liquid to form a vapor in the space
    above the liquid. See Section 12-2 in your text.

2. <u>Vapor pressure</u> is the pressure exerted in the vapor (at
    constant temperature) in dynamic equilibrium with the
    liquid state in a closed system. Dynamic equilibrium is
    established when the rate of molecules leaving the surface
    of the liquid (evaporation) is equal to the rate of the
    molecules reentering the liquid (condensation). As the
    temperature increases, the vapor pressure increases. For
    example, at $10^{\circ}C$ the vapor pressure of benzene is 45.4
    mm Hg; at $60^{\circ}C$ it is 389 mm Hg. See Section 12-3 in your
    text.

3. <u>Boiling point</u> of a liquid is the temperature at which the
    vapor pressure of the liquid is equal to the external pres-
    sure above the surface of the liquid. The normal boiling
    point of a liquid is the temperature at which the vapor
    pressure is 760 mm Hg (1 atm). To change physical states
    from liquid to gas, heat must be supplied. The quantity of
    heat required to evaporate 1 g of a given liquid at its
    boiling point at constant pressure is called the <u>heat of
    vaporization</u> of the liquid (endothermic). Heats of vapor-
    ization are normally measured at the normal boiling point
    of the liquid and under one atmosphere pressure (760 mm Hg).
    The heat of vaporization is expressed in calories per gram
    or joules per kilogram. In condensation heat is given
    off (exothermic). This heat is called the <u>heat of conden-
    sation</u> and has the same numerical value as the heat of
    vaporization. The heats of vaporization or condensation
    are given in Table 12-1 of your text.
    See Section 12-4 in your text.

4. <u>Melting point</u> is the temperature at which the kinetic
    energy of some of the particles in a solid matches the
    attractive forces in the solid and the solid begins to
    liquefy. Melting means going from the solid to the liquid.
    <u>Freezing point</u> is equal to melting point. Freezing means
    going from the liquid to the solid.

```
           melting                    ----->        = equilibrium
solid   ----------  liquid           <-----
        <---------
           freezing
```

The <u>heat</u> <u>of</u> <u>fusion</u> of a substance is the quantity of heat
required to convert 1 g of a solid substance to the liquid
state at the melting point of the substance (endothermic).
The heat of fusion is expressed in calories per gram or
joules per kilogram. The reverse of fusion is solidification. In solidification heat is given off (exothermic).
This heat is called the <u>heat</u> <u>of</u> <u>solidification</u> (crystallization) and has the same numerical value as the heat of
fusion. Heats of fusion or solidification are given in
Table 12-3 in your text. See Section 12-9 in your text.

PROBLEM 12-1

Using graph paper, plot the vapor pressure of benzene on the
vertical axis and the temperature on the horizontal axis from
the following data:

	Vapor Pressure in mm Hg		
Temperature (^{O}C)	Benzene	Chloroform	Decane
10	45.4	101	1.3
20	74.7	160	3.0
30	118	246	5.0
40	181	366	9.0
50	269	526	15
60	389	740	25
70	547	1019	39
80	754	1403	59
90	1016	--	87
100	1344	2429	126
110	1748	--	178
120	2238	3926	247
130	2825	--	337
140	3520	--	453
150	4335	--	602

175

Using this graph, answer the following:

(a) Estimate the normal boiling points (76$\overline{0}$ mm Hg) of benzene and chloroform. _____

(b) At what temperature would benzene boil if the external pressure were reduced to 50$\overline{0}$ torr?

_____

(c) Estimate the vapor pressure of decane at 125°C. _____

ANSWERS TO PROBLEM 12-1

(a) 80°C and 60°C, respectively

(b) about 67°C   Your graph should appear as follows:

(c) about 290 torr

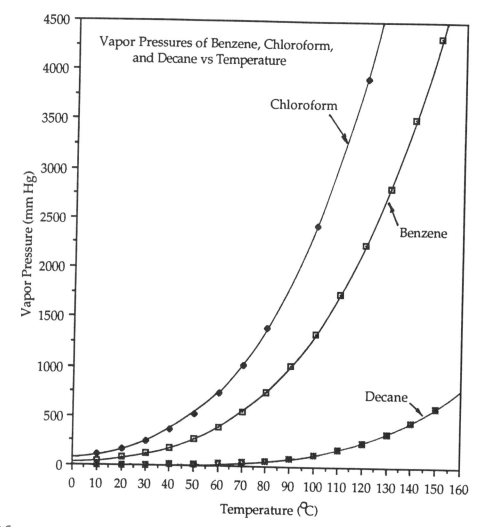

If your answers are correct, proceed to Problem 12-2. If your answers are incorrect, review Section 12-4 (Study Exercise 12-1 in your text.

Proceed to Problem 12-2.

PROBLEM 12-2

Calculate the quantity of heat energy in kilocalories required to vaporize 76.0 g alcohol (C_2H_6O) at its normal boiling point (78.3°C). See Table 12-1 in your text.

ANSWER AND SOLUTION TO PROBLEM 12-2

15.5 kcal From Table 12-1, the heat of vaporization for alcohol is 204 cal/g. Calculate the quantity of heat in kilocalories as follows:

$$76.0 \text{ g} \times \frac{204 \text{ cal}}{1 \text{ g}} \times \frac{1 \text{ kcal}}{1000 \text{ cal}} = 15.5 \text{ kcal}$$

If your answer is correct, proceed to Problem 12-3. If your answer is incorrect, review Section 12-4 (Examples 12-1 and 12-2 and Study Exercise 12-2) in your text.

Proceed to Problem 12-3.

PROBLEM 12-3

Calculate the number of kilojoules needed to vaporize 20.0 g of benzene (C_6H_6) at its normal boiling point of 80.1°C. See

Table 12-1 in your text.

ANSWER AND SOLUTION TO PROBLEM 12-3

7.88 kJ   The heat of vaporization for benzene is 3.94×10^5
J/kg. Calculate as follows:

$$20.0 \text{ g} \times \frac{1 \text{ kg}}{1000 \text{ g}} \times \frac{3.94 \times 10^5 \text{ J}}{1 \text{ kg}} \times \frac{1 \text{ kJ}}{1000 \text{ J}} = 7.88 \text{ kJ}$$

Alternate Solution

$$20.0 \text{ g} \times \frac{94.1 \text{ cal}}{1 \text{ g}} \times \frac{4.184 \text{ J}}{1 \text{ cal}} \times \frac{1 \text{ kJ}}{1000 \text{ J}} = 7.87 \text{ kJ}$$

(difference due to rounding off)

If your answer is correct, proceed to Problem 12-4. If your
answer is incorrect, re-read Section 12-4 in your text. Re-
work Problems 12-2 and 12-3.

Proceed to Problem 12-4.

PROBLEM 12-4

Calculate the melting point of ice if a pressure of 21 atm is
exerted by an ice skater on the ice.

ANSWER AND SOLUTION TO PROBLEM 12-4

-0.16°C A pressure of 1 atm will lower the melting point of
ice 0.0075°C. Therefore, the calculation is as
follows:

$$21 \text{ atm} \times \frac{0.0075 \, ^{\circ}\text{C}}{1 \text{ atm}} = 0.16^{\circ}\text{C lowering, hence the}$$
melting point of ice is -0.16°C

If your answer is correct, proceed to Problem 12-5. If your
answer is incorrect review Section 12-9 (Example 12-3 and Study
Exercies 12-3) in your text.

Proceed to Problem 12-5.

PROBLEM 12-5

Calculate the quantity of heat energy in calories evolved when
25.0 g of carbon tetrachloride (CCl_4) at -23°C is converted
from the liquid state to the solid state at -23°C. See Table
12-3 in your text.

ANSWER AND SOLUTION TO PROBLEM 12-5

127 cal   From Table 12-3, the heat of fusion for carbon tetra-
          chloride is 5.09 cal/g. Calculate the quantity of heat
          energy in calories as follows:

$$25.0 \cancel{g} \times \frac{5.09 \text{ cal}}{1 \cancel{g}} = 127 \text{ cal}$$

If your answer is correct, proceed to Problem 12-6. If your
answer is incorrect, review Section 12-9 (Examples 12-4 and
12-5 and Study Exercise 12-4) in your text.

Proceed to Problem 12-6.

PROBLEM 12-6

Calculate the quantity of heat energy in calories that must be
removed in order to convert 20.0 g of steam at 100°C to ice
at 0°C. See Tables 12-1, 12-3, and 3-4 in your text.

ANSWER AND SOLUTION TO PROBLEM 12-6

14,400 cal   Diagram the changes as follows:

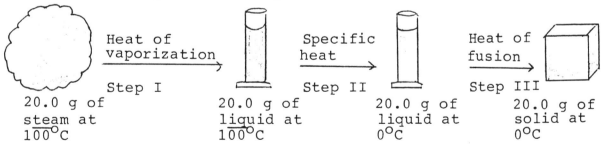

Step I	Step II	Step III
20.0 g of steam at 100°C	20.0 g of liquid at 100°C	20.0 g of liquid at 0°C

20.0 g of solid at 0°C

        From Table 12-1, the heat of vaporization of water is
        540 cal/g; hence, for the change in Step I,

$$20.0 \cancel{g} \times \frac{540 \text{ cal}}{1 \cancel{g}} = 10,800 \text{ cal}$$

179

From Table 3-4, the specific heat of water is 1.00 cal/g $^{\circ}$C; hence, for the change in Step II,

$$\frac{1.00 \text{ cal}}{1 \cancel{g} \cdot 1 \cancel{^{\circ}C}} \times 20.0 \cancel{g} \, (1\overline{00} - 0)\cancel{^{\circ}C} = 2,\overline{0}00 \text{ cal}$$

(See Section 3-5, specific heat, in your text for this calculation.)

From Table 12-3, the heat of fusion is $8\overline{0}$ cal/g; hence, for the change in Step III,

$$20.0 \cancel{g} \times \frac{8\overline{0} \text{ cal}}{1 \cancel{g}} = 1,600 \text{ cal}$$

The total heat energy removed is:
    Step I: 10,800 cal
    Step II: 2,000 cal
    Step III: 1,600 cal
Total heat energy removed = 14,400 cal

If your answer is correct, you have completed this chapter. If your answer is incorrect, review Section 12-11 (Examples 12-6, 12-7, and 12-8 and Study Exercise 12-5) in your text.

These problems conclude the chapter on liquids and solids.

Now take the sample quiz to see if you have mastered the material in this chapter.

SAMPLE QUIZ

Quiz #12

1. Calculate the quantity of heat energy in kilocalories evolved when 65.0 g of steam condense to form liquid water at its normal boiling point. (Heat of condensation = 540 cal/g)

2. Calculate the quantity of heat energy in kilojoules required to vaporize 20.7 g of alcohol at its normal boiling point of 78.3°C. (Heat of vaporization = 8.54 x 10^5 J/kg)

3. Calculate the quantity of heat energy in kilocalories required to vaporize 25.0 g of Freon-12 (CCl_2F_2) (a substance used as a refrigerant) at its normal boiling point of -29.8 $^{\circ}$C. (Heat of vaporization = 40.4 cal/g; C = 12.0 amu, Cl = 35.5 amu, F = 19.0 amu)

4. Calculate the number of kilocalories of heat energy required to convert 50.0 g of ice at 0°C to steam at exactly 100°C. (Heat of fusion = 80.0 cal/g; specific heat = 1.00 cal/g • $^{\circ}$C; heat of vaporization = 54$\overline{0}$ cal/g)

Solutions and Answers for Quiz #12

1. 65.0 g̶ x $\dfrac{54\overline{0} \text{ c̶a̶l̶}}{1 \text{ g̶}}$ x $\dfrac{1 \text{ kcal}}{1000 \text{ c̶a̶l̶}}$ = 35.1 kcal

2. 20.7 g̶ x $\dfrac{1 \text{ k̶g̶}}{1000 \text{ g̶}}$ x $\dfrac{8.54 \times 10^5 \text{ J̶}}{1 \text{ k̶g̶}}$ x $\dfrac{1 \text{ kJ}}{1000 \text{ J̶}}$ = 17.7 kJ

3. 25.0 g̶ C̶C̶l̶₂F̶₂ x $\dfrac{40.4 \text{ c̶a̶l̶}}{1 \text{ g̶}}$ x $\dfrac{1 \text{ kcal}}{1000 \text{ c̶a̶l̶}}$ = 1.01 kcal

4. 50.0 g̶ x $\dfrac{80.0 \text{ cal}}{1 \text{ g̶}}$ = 4,$\overline{0}$00 cal

$\dfrac{1.00 \text{ cal}}{1 \text{ g̶} \cdot 1 \text{ }^{\circ}\text{C̶}}$ x 50.0 g̶ x (1$\overline{0}$0 - 0)$^{\circ}$C̶ = 5,$\overline{0}$00 cal

50.0 g̶ x $\dfrac{54\overline{0} \text{ cal}}{1 \text{ g̶}}$ = 27,$\overline{0}$00 cal

Total heat energy required = 4,$\overline{0}$00 cal + 5,$\overline{0}$00 cal + 27,$\overline{0}$00 cal = 36,$\overline{0}$00 cal

36,$\overline{0}$00 c̶a̶l̶ x $\dfrac{1 \text{ kcal}}{1000 \text{ c̶a̶l̶}}$ = 36.0 kcal

181

Review Exam #4 (Chapter 10 to 12) [Answers are in ( ) to the right of each question.]

You may use the periodic table.

Element	Atomic Mass Units (amu)
Zn	65.4
Cl	35.5
H	1.0
O	16.0
C	12.0
N	14.0
K	39.1
X	2.6
Y	6.0

1. Calculate the number of grams of zinc chloride ($ZnCl_2$) that can be prepared from 32.7 g of zinc.

$$Zn(\underline{s}) + 2\ HCl(\underline{aq}) \longrightarrow ZnCl_2(\underline{aq}) + H_2(\underline{g})$$ (Balanced)

(A)

A. 68.2 g

B. 136 g

C. 272 g

D. 32.7 g

E. 0.500 g

2. Calculate the number of grams of water produced from the burning of 1.50 mol of ethane (C_2H_6).

$$2\ C_2H_6(\underline{g}) + 7\ O_2(\underline{g}) \xrightarrow{\Delta} 4\ CO_2(\underline{g}) + 6\ H_2O(\underline{g})$$ (Balanced)

(D)

A. 9.00 g

B. 54.0 g

C. 27.0 g

D. 81.0 g

E. 18.0 g

3. In the preparation of nitric acid (HNO_3), the first step is the burning of ammonia (NH_3) in oxygen (O_2) as given in the following balanced equation:

$$4 \, NH_3(g) + 5 \, O_2(g) \xrightarrow{\triangle} 4 \, NO(g) + 6 \, H_2O(g) \quad \text{(Balanced)}$$

Calculate the number of grams of nitrogen oxide (NO) that could be produced from 10.0 mol of oxygen.

A. $30\overline{0}$ g　　　　　　　　　　　　　　　　　　　　　(C)

B. $32\overline{0}$ g

C. $24\overline{0}$ g

D. 224 g

E. $15\overline{0}$ g

4. Calculate the number of liters of oxygen gas at STP that can be formed by heating 0.500 mol potassium nitrate KNO_3).

$$2 \, KNO_3(s) \xrightarrow{\triangle} 2 \, KNO_2(s) + O_2(g) \quad \text{(Balanced)}$$

A. 11.2 L　　　　　　　　　　　　　　　　　　　　　　(B)

B. 5.60 L

C. 22.4 L

D. 1.00 L

E. 0.250 L

5. Calculate the number of grams of potassium chlorate ($KClO_3$) needed to produce 2.35 liters of oxygen gas at STP.

$$2 \, KClO_3(s) \xrightarrow{\triangle} 2 \, KCl(s) + 3 \, O_2(g) \quad \text{(Balanced)}$$

A. 6.34 g　　　　　　　　　　　　　　　　　　　　　　(E)

B. 192 g

C. 19.3 g

D. 288 g

E. 8.57 g

6. Given the following unbalanced equation:

$$N_2(g) + H_2(g) \xrightarrow[\text{catalyst}]{\triangle, \text{ pressure}} NH_3(g) \quad \text{(Unbalanced)}$$

Calculate the maximum volume in liters of ammonia that could be produced from 0.060 liter of hydrogen. All reactants and products are measured at STP.

(E)

A. 0.060 L

B. 1.2 L

C. 1.8 L

D. 0.40 L

E. 0.040 L

7. A mixture containing 5.00 g hydrogen and 64.0 g oxygen is sparked so that water is formed according to the following balanced equation:

$$2 H_2(g) + O_2(g) \longrightarrow 2 H_2O(\ell) \quad \text{(Balanced)}$$

Calculate the number of grams of water that could be produced.

(B)

A. 72 g

B. 45 g

C. $45\bar{0}$ g

D. 144 g

E. 90.0 g

8. Calculate the number of moles of excess reagent remaining at the end of the reaction given in question 7.

(E)

A. 2.00 mol

B. 0.500 mol

C. 1.00 mol

D. 1.50 mol

E. 0.75 mol

184

9. Calculate the percent yield if the theoretical yield is 50.0 g and 40.0 g is actually obtained.

    A. 66.7 %   (C)

    B. 125 %

    C. 80.0 %

    D. 12.5 %

    E. 87.7 %

10. Calculate the number of kilocalories of heat energy produced by the reaction of 25.0 g oxygen with hydrogen according to the following balanced equation:

    $$2\ H_2(g)\ +\ O_2(g)\ \text{--->}\ 2\ H_2O(\ell)\ +\ 137\ \text{kcal (at } 25^{O}C)$$

    A. 107 kcal   (A)

    B. 214 kcal

    C. 53.5 kcal

    D. 175 kcal

    E. 137 kcal

11. A sample of gas has a volume of 425 mL at $25^{O}C$ and $76\overline{0}$ torr. Calculate its volume in mL at $25^{O}C$ and $63\overline{0}$ torr.

    A. 121 mL   (C)

    B. 352 mL

    C. 513 mL

    D. 829 mL

    E. 1528 mL

12. What final pressure in torr must be applied to a sample of gas having a volume of $20\overline{0}$ mL at $2\overline{0}^{O}C$ and $75\overline{0}$ torr pressure to permit the expansion of the gas to a volume of 600 mL at $2\overline{0}^{O}C$?

185

A. 117 torr   (D)

B. 75$\overline{0}$ torr

C. 2250 torr

D. 25$\overline{0}$ torr

E. 85$\overline{0}$ torr

13. A sample of gas occupies a volume of 1$\overline{00}$ mL at 1$\overline{00}$°C and 76$\overline{0}$ mm Hg. Calculate its temperature in °C if the volume is 2$\overline{00}$ mL and the pressure is 76$\overline{0}$ mm Hg.

A. 746°C   (B)

B. 473°C

C. 186°C

D. 2$\overline{00}$°C

E. 5$\overline{0}$°C

14. A certain gas occupies a volume of 5$\overline{00}$ mL at 27°C and 74$\overline{0}$ torr. Calculate its volume in mL at -73°C and 37$\overline{0}$ torr.

A. 167 mL   (B)

B. 667 mL

C. 15$\overline{0}$0 mL

D. 5$\overline{00}$ mL

E. 2$\overline{00}$0 mL

15. The volume of nitrogen, collected over water, is 19$\overline{0}$ mL at 27°C and 727 torr. Calculate the dry volume in mL of nitrogen at STP. Vapor pressure of water at 27°C is 27 torr.

A. 189 mL   (E)

B. 166 mL

C. 226 mL

D. 218 mL

E. 159 mL

186

16. Calculate the volume in liters of 0.0500 mol of oxygen gas at $30^{O}C$ and 1.10 atm.

    A.   1.50 L   (E)

    B.   31.7 L

    C.   309 L

    D.   0.120 L

    E.   1.13 L

17. Calculate the molecular mass of a gas if 0.500 g of the gas has a volume of $55\overline{0}$ mL at $27^{O}C$ and $76\overline{0}$ torr.

    A.   22.6 amu   (D)

    B.   50.0 amu

    C.   40.7 amu

    D.   22.4 amu

    E.   44.8 amu

18. Calculate the density of X_2Y gas at 4.00 atm and $27^{O}C$. (See beginning of this exam for atomic masses)

    A.   0.500 g/L   (D)

    B.   2.00 g/L

    C.   2.19 g/L

    D.   1.82 g/L

    E.   0.138 g/L

19. Calculate the volume in liters of oxygen gas produced at $27.0^{O}C$ and 1.00 atm by the heating of 0.500 mol of potassium chlorate ($KClO_3$) according to the following balanced equation:

$$2 \ KClO_3(\underline{s}) \ \xrightarrow{\triangle} \ 2 \ KCl(\underline{s}) + 3 \ O_2(\underline{g}) \quad (\underline{Balanced})$$

    A.   11.2 L   (C)

    B.   12.3 L

    C.   18.5 L

    D.   15.3 L

    E.   16.8 L

187

20. Steam will produce serious burns when it condenses on the skin. This phenomenon may be explained by the

   A. heat of condensation   (A)

   B. heat of sublimation

   C. heat of fusion

   D. heat of reaction

   E. specific heat

21. Calculate the quantity of heat in kilojoules required to vaporize 35.0 g of liquid water to steam at its normal boiling point ($10\overline{0}°C$). (Heat of vaporization = 2.26×10^6 J/kg)

   A. 7.91 kJ   (B)

   B. 79.1 kJ

   C. 7.91×10^4 kJ

   D. 7.91×10^6 kJ

   E. 7.91×10^{-5} kJ

22. Calculate the quantity of heat in calories evolved when 0.0650 mol of steam is converted to liquid water at its normal boiling point ($10\overline{0}°C$). (Heat of condensation = 540 cal/g)

   A. 632 cal   (A)

   B. 0.632 cal

   C. 11.4 cal

   D. 6.32×10^{-3} cal

   E. 6.32 cal

23. Calculate the quantity of heat in joules evolved when 27.0 g of liquid benzene is converted to solid benzene at its melting point of $6^{O}C$. (Heat of solidification =

    1.26×10^5 J/kg)

    A. 3.40×10^2 J   (D)

    B. $34\overline{0}$ J

    C. 34.0 J

    D. 3.40×10^3 J

    E. 3.40×10^6 J

24. Calculate the quantity of heat in kilocalories required to melt 29.0 g of solid carbon tetrachloride (CCl_4) to liquid carbon tetrachloride at its melting point of $-23^{O}C$. (Heat of fusion = 5.09 cal/g)

    A. 148 kcal   (E)

    B. 4.16 kcal

    C. 0.395 kcal

    D. 22.7 kcal

    E. 0.148 kcal

25. Calculate the quantity of heat in kilocalories required to convert $10\overline{0}$ g of ice at $0^{O}C$ to steam at $10\overline{0}^{O}C$. (Heat of fusion = $8\overline{0}$ cal/g, heat of vaporization = $54\overline{0}$ cal/g, specific heat of water = 1.00 cal/g$\cdot^{O}C$)

    A. 72.0 kcal   (A)

    B. 8.0 kcal

    C. 10.0 kcal

    D. 18.0 kcal

    E. 54.0 kcal

CHAPTER 13

WATER

In this chapter we will consider the physical and chemical properties of water and a related compound, hydrogen peroxide. We will also consider hydrates, crystalline salts that contain chemically bound water in definite proportions.

SELECTED TOPICS

1. A water molecule (H_2O) consists of two atoms of hydrogen and one atom of oxygen. It has a bond angle of 105°.

             Lewis structure   Structural formula

This bond angle results from the placement of the four electron pairs around the central oxygen atom (see Section 6-8, special cases of tetrahedral: bent and pyramidal, in your text). These four electron pairs consist of two pairs of bonding electrons (H-O) and two pairs of unshared electrons. This arrangement results in a tetrahedron with the molecule considered bent. See Section 13-2 in your text.

2. The oxygen atom is more electronegative than the hydrogen atom (see Section 6-4, electronegativities in covalent bonds, in your text). This difference in electronegativities creates an unequal sharing of electrons between oxygen and hydrogen with the oxygen atom being relatively negative

(δ^-) and the hydrogen atom relatively positive (δ^+).

This unequal sharing of electrons produces a polar bond or polar covalent bond. A <u>polar bond</u> (polar covalent bond) is a type of chemical bond formed by the unequal sharing of electrons between two atoms whose electronegativities differ.

$$\delta^-$$

$$O - H \quad \delta^+$$

$$\delta^+ \quad H$$

Water molecules have a net dipole moment. A molecule having a net dipole moment is one in which the centers of positive and negative charges do not coincide but are separated by a finite distance.

$$\begin{array}{c} O\!\!-\!\!H \\ /\;\nwarrow \\ H\quad X \end{array}$$

The arrow symbol (+——>) indicates this moment with the head of the arrow pointing to the negative center. If the bond angle were 180° (linear), then no net dipole moment would exist for the compound, because the center of positive charges would coincide with the center of negative charge on the oxygen atom.

$$H\!\!-\!\!O\!\!-\!\!H \qquad \text{dipole moment} = 0$$
$$180^O \quad \text{(linear)}$$

Therefore, the shape of a molecule is a very important factor in determining dipole moment. See Section 13-3 in your text.

3. Water exhibits some extraordinary properties when compared to hydrogen compounds (H_2S, H_2Se, H_2Te) of other elements in the oxygen group (group VIA, 16). These properties are high melting point, high boiling point, high heat of fusion (kcal/mol and kJ/mol) and high heat of vaporization (kcal/mol and kJ/mol). To explain these high values (see Table 13-2 in your text), we use hydrogen bonding. As we previously mentioned, water is a polar compound with a net dipole moment. The partial negative charge, δ^-, on the

oxygen atom from the unshared pairs of electrons, in one water molecule attract a partial positive charge, δ^+,

on the hydrogen atom of another water molecule. This attraction forms a weak linkage called a hydrogen bond. A hydrogen bond is a type of weak chemical bond formed when a hydrogen atom bonded to a highly electronegative atom, (X), also bonds to another electronegative atom. These electronegative atoms are F, O, and N. This hydrogen bond is shown with a broken line (----) as follows:

$$X\!\!-\!\!H \;----\; X\!\!-\!\!H \qquad\qquad \begin{array}{c} O\!\!-\!\!H \;----\; O\!\!-\!\!H \\ / \qquad\qquad\quad / \\ H \qquad\qquad\quad H \end{array}$$

            General case   Water

The hydrogen bond is much weaker than a covalent bond. The average hydrogen bond energy is about 5 kcal/mol; the average covalent bond energy is about 120 kcal/mol. Heat energy is necessary to break some of the attractive forces in the solid or liquid in the transformation from solid to liquid to gas. With water these attractive forces include hydrogen bonding. Therefore more energy is necessary and these values (see Table 13-2 in your text) are higher than for nonhydrogen-bonded compounds. The hydrogen compound of the other elements in group VIA (16) [H_2S, H_2Se, and H_2Te] do not hydrogen bond, because S, Se, and Te are not as electronegative as O. See Section 13-4 in your text.

4. The following are some important reactions in which water is produced:

(a) combustion: $2 C_2H_6(g) + 7 O_2(g) \xrightarrow{\triangle} 4 CO_2(g) + 6 H_2O(g)$

        ethane

Heat energy is also released.

(b) combination of hydrogen and oxygen:

$2 H_2(g) + O_2(g) \xrightarrow{\text{"spark"}} 2 H_2O(g)$

Heat energy is also released.

(c) product of neutralization reactions:

$NaOH(aq) + HCl(aq) \longrightarrow NaCl(aq) + H_2O(\ell)$

The following are some reactions of water:

(a) electrolysis: $2 H_2O(\ell) \xrightarrow[\text{electric current}]{\text{direct}} 2 H_2(g) + O_2(g)$

(b) reactions with certain metals:

The first five metals (Li, K, Ba, Ca, and Na) in the electromotive or activity series react with water to form the metal hydroxide and hydrogen gas. Remember to write hydrogen gas as a diatomic molecule.

$2 Na(s) + 2 HOH(\ell) \longrightarrow 2 NaOH(aq) + H_2(g)$

The next five metals (Mg, Al, Zn, Fe, and Cd) in the series react with water in the form of steam to form the metal oxide and hydrogen gas.

$2 Al(s) + 3 H_2O(g) \xrightarrow{\triangle} Al_2O_3(s) + 3 H_2(g)$

The other metals in the electromotive or activity series do not react with water. See Sections 13-5 and 13-6 in your text.

5. <u>Hydrates</u> are crystalline compounds that contain chemically bound water in definite proportions. An example of a hydrate is Na_2SO_4 10 H_2O, sodium sulfate decahydrate (see Table 7-1 in your text for a review of the Greek prefixes). We can calculate the percent of water in a hydrate by determining the formula mass of the hydrate and dividing this value in amu into the total contribution of the water in amu and multiplying by 100. We can determine the formula of the hydrate from the percent of water in the hydrate. First, we convert the percent of water and the percent of anhydrous salt to grams of water and anhydrous salt, and determine respectively, the moles of water and anhydrous salt. Second, we determine the ratio of the anhydrous salt to the water by dividing by the smallest value of moles. This ratio in whole numbers gives us the number of water molecules in the hydrate. See Section 13-7 in your text.

6. Hydrogen peroxide has the formula H_2O_2. It is prepared in the laboratory by treating barium peroxide (BaO_2) with an aqueous solution of sulfuric acid as follows:

$$BaO_2 (\underline{s}) + H_2SO_4 (\underline{aq}) \longrightarrow BaSO_4 (\underline{s}) + H_2O_2 (\underline{aq})$$

Hydrogen peroxide readily decomposes to water and oxygen with heat, light, or a catalyst as follows:

$$2\ H_2O_2 (\underline{aq}) \ \xrightarrow[\text{or catalyst}]{\triangle,\ \text{light}} \ 2\ H_2O (\ell) + O_2 (\underline{g})$$

The catalyst can be a substance such as silver, carbon, manganese(IV) oxide, saliva, dirt, or an enzyme in the blood. See Section 13-9 in your text.

PROBLEM 13-1

Represent the net dipole moment; if any, for the following molecules. (<u>Hint</u>: Write the Lewis structure and structural formula (see Section 6-7, in your text), consider the electronegativities of the element (see Figure 6-11), then decide on the shape of the molecule, that is linear, trigonal planar, tetrahedral, bent, or pyramidal (see Section 6-8).

(a) HBr

(a) _____

(b) CS_2

(b) _____

ANSWERS AND SOLUTIONS TO PROBLEM 13-1

+—>
H Br   Bromine is more electronegative than hydrogen (see
        Figure 6-11 in your text for the order of electro-
        negativities). Therefore, the arrow is pointing to the
        Br atom.

dipole   The electron-dot formula and structural formula for
moment CS_2 are as follows: 
= 0

$$\ddot{\underset{\cdot\cdot}{S}} \overset{\times}{\underset{\times}{\circ}} C \overset{\cdot\cdot}{\underset{\times}{\circ}} \ddot{\underset{\cdot\cdot}{S}}$$ S══C══S

        The molecule is linear; therefore, the bond angle is
        0.

If your answers are correct, proceed to Problem 13-2. If your
answers are incorrect, review Section 13-3 (Study Exercise 13-
1) in your text.

Proceed to Problem 13-2.

PROBLEM 13-2

Explain the apparent irregularities in the series of melting
points shown for the compounds of HF, HCl, HBr, and HI. The
electronegativities are as follows: F = 4.0, Cl = 3.0, Br =
2.8, and I = 2.5.

Compound	Mp $^{\circ}C$
HF	– 83.7
HCl	– 114.2
HBr	– 86.9

_____

_____

_____

_____

_____

_____

_____

ANSWER AND SOLUTION TO PROBLEM 13-2
The melting point of HF is higher than that of HCl and HBr.
Because of the high electronegativity of F, HF forms hydrogen
bonds (H-F---H-F). Cl and Br have low electronegativities and
do not hydrogen bond. The energy required to separate
molecules of HF include energy to break the hydrogen bonds.
The HCl and HBr do not have hydrogen bonds to break and thus
separate with less energy needed.

If your answer is correct, proceed to Problem 13-4. If your
answer is incorrect, review Section 13-2 (Study Exercise 13-2)
in your text.

Proceed to Problem 13-3.

PROBLEM 13-3

Complete and balance the following chemical reaction equations;
indicate any precipitate by (s) and any gas by (g): (You may
use the periodic table, the electromotive (activity) series,
and the rules for the solubility of inorganic substances in
water.)

(a) $C_3H_8(g)$ + $O_2(g)$ $\overset{\triangle}{===}$>
    propane

(b) Ca(s) + HOH(ℓ) --->

(c) Cd(s) + $H_2O(g)$ $\overset{\triangle}{==}$>

(d) CaO(s) + HCl(aq) --->

ANSWERS AND SOLUTIONS TO PROBLEM 13-3

(a) $C_3H_8(g)$ + 5 $O_2(g)$ $\overset{\triangle}{===}$> 3 $CO_2(g)$ + 4 $H_2O(g)$

    This is a combustion reaction yielding CO_2 and H_2O as pro-
    ducts. Heat energy is also released.

(b) Ca(s) + 2 HOH(ℓ) ---> Ca(OH)$_2$(s) + $H_2(g)$

    The first five elements in the electromotive or activity
    series react with water to form the metal hydroxide and
    hydrogen gas.

(c) Cd(s) + $H_2O(g)$ $\overset{\triangle}{===}$> CdO(s) + $H_2(g)$

    The next five elements in the electromotive or activity
    series react with steam to form the metal oxide and hydro-
    gen gas.

(d) CaO(s) + 2 HCl(aq) ---> CaCl$_2$(aq) + $H_2O(\ell)$

    This is a neutralization reaction involving a metal oxide
    (basic oxide) and an acid. Water is one of the products.

If your answers are correct, proceed to Problem 13-4. If your answers are incorrect, review Sections 13-5 and 13-6 (Study Exercise 13-3) in your text.

Proceed to Problem 13-4.

PROBLEM 13-4

Calculate the percent of water in the following hydrates:

(a) sodium sulfate decahydrate

_____

(b) copper(II) chlorate hexahydrate

_____

ANSWERS AND SOLUTIONS TO PROBLEM 13-4

(a) 55.9 %   The formula for sodium sulfate decahydrate is $Na_2SO_4 \cdot 10\ H_2O$. The formula mass of $Na_2SO_4 \cdot 10\ H_2O$ is 322.1 amu (Use the Table of Approximate Atomic Masses in the back of your text.)

$2 \times 23.0 = 46.0$ amu
$1 \times 32.1 = 32.1$ amu
$4 \times 16.0 = 64.0$ amu
$10 \times 18.0 = \underline{180.0}$ amu (molecular mass H_2O = 18.0 amu)
Formula mass of $= 322.1$ amu
$Na_2SO_4 \cdot 10\ H_2O$

$$\frac{180.0\ \cancel{amu}}{322.1\ \cancel{amu}} \times 100 = 55.9\ \%\ water$$

(b) 31.9 %   The formula for copper(II) chlorate hexahydrate is $Cu(ClO_3)_2 \cdot 6\ H_2O$. The formula mass of $Cu(ClO_3)_2 \cdot 6\ H_2O$ is 338.5 amu.

$1 \times 63.5 = 63.5$ amu
$2 \times 35.5 = 71.0$ amu
$6 \times 16.0 = 96.0$ amu
$6 \times 18.0 = \underline{108.0}$ amu (molecular mass H_2O = 18.0 amu)
Formula mass of $= 338.5$ amu
$Cu(ClO_3)_2 \cdot 6\ H_2O$

$$\frac{108.0 \; \cancel{amu}}{338.5 \; \cancel{amu}} \times 100 = 31.9 \; \%$$

● If your answers are correct, proceed to Problem 13-5. If your answers are incorrect, review Section 13-7 (Example 13-1 and Study Exercise 13-4) in your text.

Proceed to Problem 13-5.

PROBLEM 13-5

Calculate the formula for the following hydrates:

(a)   lithium chloride, containing 29.8 percent water

_____

(b)   potassium carbonate, containing 20.7 percent water

_____

ANSWERS AND SOLUTIONS TO PROBLEM 13-5

(a) $LiCl \cdot H_2O$ If 29.8 percent of the hydrate is water, 70.2 percent (100.0 - 29.8 = 70.2) must be lithium chloride. Hence, in 100 g of the hydrate, there would be 29.8 g of water and 70.2 g of lithium chloride. Therefore, the first step is to calculate the number of moles of water and of lithium chloride in 100 g of the hydrate, as follows:

$$29.8 \; \cancel{g \; H_2O} \times \frac{1 \; mol \; H_2O}{18.0 \; \cancel{g \; H_2O}} = 1.66 \; mol \; H_2O$$

$$70.2 \; \cancel{g \; LiCl} \times \frac{1 \; mol \; LiCl}{42.4 \; \cancel{g \; LiCl}} = 1.66 \; mol \; LiCl$$

(Molar mass of LiCl = 42.4 g)

The second step is to express these relationships in small whole numbers by dividing by the smallest value, as follows:

$$For \; LiCl = \frac{1.66}{1.66} = 1.00$$

$$For \; H_2O = \frac{1.66}{1.66} = 1.00$$

Hence, the formula for the hydrate is $LiCl \cdot H_2O$

197

(b) $K_2CO_3 \bullet 2H_2O$ If 20.7 percent of the hydrate is water, 79.3 percent (100.0 - 20.7 = 79.3) must be potassium carbonate. Hence, in 100 g of the hydrate there would be 20.7 g of water and 79.3 g of potassium carbonate. Next, calculate the moles of each.

$$20.7 \text{ g } \cancel{H_2O} \times \frac{1 \text{ mol } H_2O}{18.0 \cancel{\text{ g } H_2O}} = 1.15 \text{ mol } H_2O$$

$$79.3 \text{ g } \cancel{K_2CO_3} \times \frac{1 \text{ mol } K_2CO_3}{138.2 \cancel{\text{ g } K_2CO_3}} = 0.574 \text{ mol } K_2CO_3$$

(Molar mass of K_2CO_3 = 138.2 g)

Divide by the smallest value.

For K_2CO_3 = $\frac{0.574}{0.574}$ = 1.00

For H_2O = $\frac{1.15}{0.574}$ = 2.00

The formula for the hydrate is $K_2CO_3 \bullet 2H_2O$

If your answers are correct, proceed to Problem 13-6. If your answers are incorrect, review Section 13-7 (Example 13-2 and) Study Exercise 13-5) in your text.

Proceed to Problem 13-6.

PROBLEM 13-6

Complete and balance the following chemical reaction equations; indicate any precipitate by (s) and any gas by (g): (You may use the periodic table and the rules for the solubility of inorganic substances in water.)

(a) $BaO_2 (\underline{s}) + H_2SO_4 (\underline{aq})$ --->

(b) $H_2O_2 (\underline{aq})$ $\xrightarrow{\text{Ag}}$

ANSWERS AND SOLUTIONS TO PROBLEM 13-6

(a) $BaO_2 (\underline{s}) + H_2SO_4 (\underline{aq})$ ---> $BaSO_4 (\underline{s}) + H_2O_2 (\underline{aq})$

This is the laboratory preparation for hydrogen peroxide.

(b) $2 H_2O_2 (\underline{aq})$ $\xrightarrow{\text{Ag}}$ $2 H_2O (\ell) + O_2 (\underline{g})$

Ag catalyzes the decomposition.

If your answers are correct, you have completed this chapter. If your answers are incorrect, review Section 13-9 (Study Exercise 13-6) in your text.

These problems conclude the chapter on water.

Now take the sample quiz to see if you have mastered the material in this chapter.

SAMPLE QUIZ

<u>Quiz #13</u>

You may use the periodic table, the electromotive (activity) series, and the rules for the solubility of inorganic substances in water.

1. Explain the apparent irregularity in the series of boiling points shown for the compounds NH_3, PH_3, and AsH_3. The

    electronegativities are as follows: N = 3.0, P = 2.1, and As = 2.1.

Compound	Bp $^{\circ}$C (1 atm)
NH_3	-33.4
PH_3	-87.7
AsH_3	-55

2. Complete and balance the following chemical reaction equations; indicate any precipitate by (<u>s</u>) and any gas by (<u>g</u>):

    (a) BaO_2 (<u>s</u>) + H_2SO_4 (<u>aq</u>) --->

    (b) CO_2 (<u>g</u>) + $Ca(OH)_2$ (<u>aq</u>) --->

    (c) SrO (<u>s</u>) + HCl (<u>aq</u>) --->

    (d) Al (<u>s</u>) + H_2O (<u>g</u>) $\xrightarrow{\triangle}$

3. Calculate the percent of water in $AlCl_3$ $6H_2O$ (aluminum chloride hexahydrate). $Al = 27.0$ amu, $Cl = 35.5$ amu, $H = 1.0$ amu, $O = 16.0$ amu

4. Calculate the formula of a hydrate of sodium carbonate (Na_2CO_3), containing 63.0 percent water. $Na = 23.0$ amu, $C = 12.0$ amu, $O = 16.0$ amu, $H = 1.0$ amu

Answers and Solutions to Quiz #13

1. The melting point of NH_3 is higher than that of PH_3 and AsH_3. Because of the high electronegativity of N, NH_3 forms hydrogen bonds ($H_2N-H---NH_3$). P and As have low electronegativities and do not hydrogen bond. The energy required to separate molecules of HF include energy to break the hydrogen bonds. The PH_3 and AsH_3 do not have hydrogen bonds to break and thus separate with less energy needed.

2. (a) $BaO_2(s) + H_2SO_4(aq) \longrightarrow BaSO_4(s) + H_2O_2(aq)$

   (b) $CO_2(g) + Ca(OH)_2(aq) \longrightarrow CaCO_3(s) + H_2O(\ell)$

   (c) $SrO(s) + 2 HCl(aq) \longrightarrow SrCl_2(aq) + H_2O(\ell)$

   (d) $2 Al(s) + 3 H_2O(g) \overset{\triangle}{\longrightarrow} Al_2O_3(s) + 3 H_2(g)$

3. 1 x 27.0 = 27.0 amu $\dfrac{108.0 \text{ amu}}{241.5 \text{ amu}}$ x 100 = 44.7 %
    3 x 35.5 = 106.5 amu
    6 x 18.0 = 108.0 amu
    FM   = 241.5 amu

200

4. $63.0 \text{ g } H_2O \times \dfrac{1 \text{ mol } H_2O}{18.0 \text{ g } H_2O} = 3.50 \text{ mol } H_2O$

$37.0 \text{ g } Na_2CO_3 \times \dfrac{1 \text{ mol } Na_2CO_3}{106.0 \text{ g } Na_2CO_3} = 0.349 \text{ mol } Na_2CO_3$

For water $\dfrac{3.50}{0.349} \doteq 10 \qquad Na_2CO_3 \quad 10 \; H_2O$

(Molar mass of $Na_2CO_3 = 106.0$ g)

CHAPTER 14

SOLUTIONS AND COLLOIDS

In this chapter we will consider solutions. Solutions are homogeneous throughout. They are composed of two or more substances and their composition can usually be varied within certain limits. Colloids are intermediate between matter in solution and matter in suspension. In a solution the particles are homogeneously dispersed and do not settle out on standing because they are partially bound to solvent molecules. In a suspension the particles are not bound to the solvent particles and do settle out on standing. In colloids the particles are not appreciably bound to the solvent molecules, but do not settle out on standing.

SELECTED TOPICS

1. A solution is composed of a solute and a solvent. The solute is the component in lesser quantity; the solvent is the component in greater quantity. In a 5.00 percent glucose (dextrose) solution in water, the glucose is the solute and the water is the solvent. See Section 14-1 in your text.

2. Henry's law states that the solubility of a gas in a liquid is directly proportional to the partial pressure of the gas above the liquid. For example, if the partial pressure of the gas above the liquid is tripled, the solubility of the gas in the liquid is also tripled. See Section 14-3, pressure, in your text.

3. A saturated solution is a solution that contains just as much solute as can be dissolved in the solvent by ordinary means. Dissolved solute is in dynamic equilibrium ($\rightleftharpoons$,

   see Section 12-3, in your text) with any undissolved solute. That is, the rate of dissolution (dissolving) of an undissolved solute is exactly equal to the rate of crystallization of dissolved solute:

$$\text{undissolved solute} \underset{\text{rate of crystallization}}{\overset{\text{rate of dissolution}}{\rightleftharpoons}} \text{dissolved solute}$$

   An unsaturated solution is a solution in which the concentration of solute is less than that of the saturated solution under the same conditions. A supersaturated solution is a solution in which the concentration of solute is greater than that possible in a saturated solution under the same conditions. A supersaturated solution is unstable and will revert to a saturated solution if a "seed" crystal of the solute is added. At 25°C, 36 g of sodium chloride dissolves in 100 g of water to form a saturated solution.

A solution formed with less than 36 g of sodium chloride in 100 g of water would be considered an unsaturated solution. A solution containing more than 36 g of sodium chloride in 100 g of water would be considered a supersaturated solution. See Section 14-4 in your text.

4. In percent by mass the concentration of a solution is expressed as the parts by mass of solute per 100 parts by mass of solution.

$$\% \text{ by mass} = \frac{\text{mass of solute}}{\text{mass of solution}} \times 100$$

For example, a 10.0 percent aqueous sodium chloride solution would contain 10.0 g of sodium chloride in 100 g of solution (90.0 g of water). See Section 14-6 in your text.

5. In parts per million (ppm) the concentration of a solution is expressed as parts by mass of solute per 1,000,000 parts by mass of solution. These solutions are very dilute; therefore, the density of the solution is assumed to be 1.00 g/mL.

$$\text{parts per million (ppm)} = \frac{\text{mass of solute}}{\text{mass of solution}} \times 1,000,000$$

A 6.0 parts per million aqueous sodium chloride solution would contain 6.0 g of sodium chloride in 1,000,000 g of solution. See Section 14-7 in your text.

6. Molarity (M) is the concentration of a solution expressed as the number of moles of solute per liter of solution.

$$M = \text{molarity} = \frac{\text{moles of solute}}{\text{liter of solution}}$$

A 1.00 molar sodium chloride solution would contain 1.00 mol (58.5 g) of sodium chloride dissolved in enough water to make the volume of the solution equal to one liter.

Concentrated (molar) solutions are called stock solutions. These solutions are used to prepare diluted solutions by adding water to the stock solution. In calculations involving these stock solutions or diluted solutions, the number of moles of solute in either the stock solution or the diluted solution must be calculated. See Section 14-8 in your text.

7. Normality (N) is the concentration of a solution expressed as the number of equivalents of solute per liter of solution.

$$N = \text{normality} = \frac{\text{equivalents of solute}}{\text{liter of solution}}$$

One equivalent (abbreviated eq) of any acid is the quantity of a substance that reacts to yield 1 mol [6.02×10^{23} (Avogadro's number)] of hydrogen ions (H^+). One equivalent of any base is the quantity of a substance that reacts with 1 mol (6.02×10^{23}) of hydrogen ions or supplies 1 mol (6.02×10^{23}) of hydroxide ions (OH^-). One equivalent of any acid will combine exactly with one equivalent of any base.

The equivalent mass in grams (one equivalent) of an acid is determined by dividing the molar mass of the acid by the number of moles of hydrogen ions per mole of acid used in the reaction. The equivalent mass in grams (one equivalent) of a base is determined by dividing the molar mass of the base by the number of moles of hydrogen ions that combine with 1 mole of the base or the number of moles of hydroxide ions per mole of base used in the reaction. In all cases, we must consider the reaction.

EQUIVALENT MASSES OF SOME ACIDS OR BASES

Acid or Base	1 mol	1 eq
HCl	36.5 g	36.5 g/1 = 36.5 g
H_2SO_4	98.1 g	98.1 g/2 = 49.0 g (Both H^+ replaced)
H_2SO_4	98.1 g	98.1 g/1 = 98.1 g (One H^+ replaced)
$NaOH$	40.0 g	40.0 g/1 = 40.0 g
$Ca(OH)_2$	74.1 g	74.1 g/2 = 37.0 g (Both OH^- replaced)

A 1.00 $\underline{N}$ aqueous sulfuric acid solution in which both hydrogen ions are replaced would contain 1.00 eq (98.1 g/2 = 49.0 g) of sulfuric acid dissolved in enough water to make the volume of the solution equal to one liter.

Molarity and normality are similar except in molarity we use moles and in normality we use equivalents. Normality is always a multiple of molarity, because there are 1, 2, 3, etc. equivalents per mole. A one molar sulfuric acid (H_2SO_4) solution would be a two normal solution in which both hydrogen ions are replaced because there are two equivalents per mole of sulfuric acid. See Section 14-9 in your text.

8. $\underline{\text{Molality}}$ ($\underline{m}$) is the concentration of a solution expressed as the number of moles of solute per kilogram of solvent.

$$\underline{m} = \text{molality} = \frac{\text{moles of solute}}{\text{kilogram of solvent}}$$

A 1.00 m aqueous sodium chloride solution would contain 1.00 mol (58.5 g) of sodium chloride dissolved in 1.00 kilogram (1000 g) of water. See Section 14-10 in your text.

9. Colligative properties of solutions are properties that depend only on the number of particles of solute present in the solution and not on the actual identity of the particles. These colligative properties are (1) boiling-point elevation, and (2) freezing-point depression. In this discussion we will consider only the effect of nonvolatile and nonionized solute particles on the volatile solvent because volatile and ionized solute particles have a more complicated effect on the solvent. We calculate the boiling-point elevation using the following equation:

$$\triangle T_b = \text{molality (m)} \times K_b$$

with $\triangle T_b$ being the boiling-point elevation and K_b, the boiling-point-elevation constant for the volatile solvent. We calculate the freezing-point depression using the following equation:

$$\triangle T_f = \text{molality (m)} \times K_f$$

with the $\triangle T_f$ being the freezing-point lowering and K_f the freezing-point-depression constant for the solvent. A 1.00 m aqueous solution of a nonvolatile solute such as sucrose (sugar) or urea will raise the normal boiling point (at 1.00 atm) of water by 0.52°C from 100.00°C to 100.52°C and will lower the freezing point of water by 1.86°C from 0.00°C to -1.86°C.

We can also use the boiling-point elevation and freezing-point depression to calculate the molecular mass and molar mass of a compound. Using the boiling-point or freezing-point of a solution of the unknown, we can calculate the molality of the unknown solution from the respective equations.

$$\text{molality (m)} = \frac{\triangle T_b}{Kb} \qquad \text{or molality (m)} = \frac{\triangle T_f}{K_f}$$

From the amount in grams of the unknown dissolved in a given amount of solvent in grams and the molality of the solution (calculated as given above), we can calculate the molecular mass and molar mass of the unknown solving for the units - grams per mole which are equal to the molecular mass in amu and the molar mass in grams. See Section 14-12 in your text.

PROBLEM 14-1

Using graph paper, plot the solubility of potassium sulfate (K_2SO_4) on the vertical axis and the temperature on the horizontal axis from the following data:

Temperature in °C	Solubility of K_2SO_4 in g/100 g H_2O
0	7.4
20	11.1
40	14.8
60	18.2
80	21.4
100	24.1

Using this graph, determine the solubility of potassium sulfate in grams per 100 g of water at the following temperatures:

(a)   30°C   _____

(b)   70°C   _____

(c)   45°C   _____

ANSWERS TO PROBLEM 14-1

(a)   13.0 g/100 g H_2O

(b)   19.8 g/100 g H_2O

(c)   15.6 g/100 g H_2O

Your graph should appear as follows:

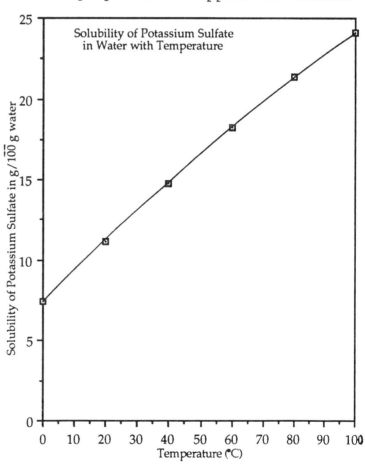

206

If your answers are correct, proceed to Problem 14-2. If your answers are incorrect, review Section 14-3, temperature (Study Exercise 14-1), in your text.

Proceed to Problem 14-2.

PROBLEM 14-2

Calculate the solubility in grams per liter of a certain gas in water at a partial pressure of 2.45 atm and $0^\circ C$. The solubility is 0.256 grams per liter at a total pressure of 1.000 atm and $0^\circ C$. (See Appendix V in your text for the vapor pressure of water.)

_____

ANSWER AND SOLUTION TO PROBLEM 14-2

0.631 g/L The total pressure is the sum of the partial pressures of water and the gas, according to Dalton's law of partial pressures.

$$P_{total} = P_{gas} + P_{water}$$

P_{water} at $0^\circ C$ = 0.0061 atm (see Appendix V in your text)

P_{gas} = 1.000 atm - 0.0061 atm = 0.994 atm

Solubility$_{old}$ = 0.256 g/L P_{old} = 0.994 atm ⦙ pressure
⦙ increases;
Solubility$_{new}$ = ? P_{new} = 2.45 atm ⦙ solubility
↓ increases

Solubility$_{new}$ = solubility$_{old}$ x pressure factor

Solubility$_{new}$ = 0.256 g/L x $\dfrac{2.45 \; \text{atm}}{0.994 \; \text{atm}}$ = 0.631 g/L

If your answer is correct, proceed to Problem 14-3. If your answer is incorrect, review Section 14-3, pressure (Example 14-1 and Study Exercise 14-2) in your text.

Proceed to Problem 14-3.

PROBLEM 14-3

Calculate the percent by mass of the solute in each of the following aqueous solutions:

(a)   4.75 g of sodium chloride in 87.0 g of solution

(b)   9.80 g of sodium chloride in 110.0 g of water

(a) _____

(b) _____

ANSWERS AND SOLUTIONS TO PROBLEM 14-3

(a)   5.46 %   The total mass of the solution is 87.0 g; therefore calculate the percent of sodium chloride as follows:

$$\frac{4.75 \text{ g NaCl}}{87.0 \text{ g solution}} \times 100 = 5.46 \text{ \% NaCl}$$

(b)   8.18 %   The total mass of the solution is 119.8 g (9.80 g NaCl + 110.0 g water). Calculate the percent sodium chloride as follows:

$$\frac{9.80 \text{ g NaCl}}{119.8 \text{ g solution}} \times 100 = 8.18 \text{ \% NaCl}$$

If your answers are correct, proceed to Problem 14-4. If your answers are incorrect, review Section 14-6 (Examples 14-2 and 14-3 and Study Exercise 14-3) in your text.

Proceed to Problem 14-4.

PROBLEM 14-4

Calculate the number of grams of solute that must be dissolved in

(a) 125 g of water in the preparation of a 12.0 percent sodium chloride solution

(b) 28$\overline{0}$ g of water in the preparation of a 25.0 percent sodium chloride solution

(a) _____

(b) _____

ANSWERS AND SOLUTIONS TO PROBLEM 14-4

(a) 17.0 g In this solution there would be 12.0 g of sodium chloride for every 88.0 g of water (100.0 g solution - 12.0 g sodium chloride = 88.0 g water). Calculate the number of grams of sodium chloride needed for 125 g of water as follows:

$$125 \text{ g } H_2O \times \frac{12.0 \text{ g NaCl}}{88.0 \text{ g } H_2O} = 17.0 \text{ g NaCl}$$

(b) 93.3 g In this solution there would be 25.0 g of sodium chloride for every 75.0 g of water (100.0 g solution - 25.0 g sodium chloride = 75.0 g water). Calculate as follows:

$$28\overline{0} \text{ g } H_2O \times \frac{25.0 \text{ g NaCl}}{75.0 \text{ g } H_2O} = 93.3 \text{ g NaCl}$$

If your answers are correct, proceed to Problem 14-5. If your answers are incorrect, review Section 14-6 (Example 14-4) in your text.

Proceed to Problem 14-5.

PROBLEM 14-5

Calculate the number of grams of water that must be added to

(a) 12.0 g sodium chloride in the preparation of a 18.0 percent sodium chloride solution

(b) 50.0 g sodium chloride in the preparation of a 5.00 percent sodium chloride solution

(a) _____

(b) _____

ANSWERS AND SOLUTIONS TO PROBLEM 14-5

(a) 54.7 g In a 18.0 percent sodium chloride solution, there
are 18.0 g of sodium chloride for every 82.0 g
water (100.0 g solution - 18.0 g sodium chloride =
82.0 g water). Calculate the number of grams of
water needed for 12.0 g sodium chloride as follows:

$$12.0 \text{ g NaCl} \times \frac{82.0 \text{ g H}_2\text{O}}{18.0 \text{ g NaCl}} = 54.7 \text{ g H}_2\text{O}$$

(b) $95\overline{0}$ g In a 5.00 percent sodium chloride solution, there
are 5.00 g of sodium chloride for 95.0 g of water
(100.0 g solution - 5.00 g sodium chloride = 95.0
g water). Calculate the number of grams of water
needed for 50.0 g sodium chloride as follows:

$$50.0 \text{ g NaCl} \times \frac{95.0 \text{ g H}_2\text{O}}{5.00 \text{ g NaCl}} = 95\overline{0} \text{ g H}_2\text{O}$$

If your answers are correct, proceed to Problem 14-6. If your
answers are incorrect, review Section 14-6 (Example 14-5 and
Study Exercise 14-4) in your text.

Proceed to Problem 14-6.

PROBLEM 14-6

Calculate the number of grams of solution necessary to provide the following:

(a)   9.55 g of sodium chloride from a 15.0 percent aqueous sodium chloride solution

(b)   25.6 g of sodium chloride from a 12.0 percent aqueous sodium chloride solution

(a) _____

(b) _____

ANSWERS AND SOLUTIONS TO PROBLEM 14-6

(a)   63.7 g   In a 15.0 percent sodium chloride solution, there are 15.0 g of sodium chloride in 100.0 g solution. Calculate the number of grams of solution needed for 9.55 g of sodium chloride as follows:

$$9.55 \; \cancel{g \; NaCl} \times \frac{100.0 \; g \; solution}{15.0 \; \cancel{g \; NaCl}} = 63.7 \; g \; solution$$

(b)   213 g   In a 12.0 percent sodium chloride solution, there are 12.0 g sodium chloride in 100.0 g of solution. Calculate the number of grams of solution needed for 25.6 g of sodium chloride as follows:

$$25.6 \; \cancel{g \; NaCl} \times \frac{100.0 \; g \; solution}{12.0 \; \cancel{g \; NaCl}} = 213 \; g \; solution$$

If your answers are correct, proceed to Problem 14-7. If your answers are incorrect, review Section 14-6 (Example 14-6) in your text.

Proceed to Problem 14-7.

211

PROBLEM 14-7

Calculate the parts per million (ppm) of the solute in each of the following aqueous solutions. Assume that the density of the very dilute water sample is 1.00 g/mL.

(a)   185 mg of sodium (Na^+) ions in $85\overline{0}$ mL of a water sample

(b)   0.250 mg copper(II)(Cu^{2+}) ions in $75\overline{0}$ mL of a water sample. This concentration in parts per million is the killing dosage of copper(II) ions for ridding a lake of carp.

(a) _____

(b) _____

ANSWERS AND SOLUTIONS TO PROBLEM 14-7

(a)   218 ppm $\dfrac{185 \text{ mg Na}^+}{85\overline{0} \text{ mL water sample}}$ x $\dfrac{1 \text{ mL sample}}{1.00 \text{ g sample}}$

x $\dfrac{1 \text{ g sample}}{100\overline{0} \text{ mg sample}}$ x 1,000,000 = 218 ppm

(b)   0.333 ppm $\dfrac{0.250 \text{ mg Cu}^{2+}}{75\overline{0} \text{ mL water sample}}$ x $\dfrac{1 \text{ mL sample}}{1.00 \text{ g sample}}$

x $\dfrac{1 \text{ g sample}}{100\overline{0} \text{ mg sample}}$ x 1,000,000 = 0.333 ppm

If your answers are correct, proceed to Problem 14-8. If your answers are incorrect, review Section 14-7 (Example 14-7 and Study Exercise 14-5) in your text.

Proceed to problem 14-8.

PROBLEM 14-8

Calculate the number of milligrams of solute dissolved in the following: (Assume that the density of the very dilute water sample is 1.00 g/mL.)

(a) 1.20 L of a water sample having 81 ppm sulfate (SO_4^{2-}) ions

(b) 6.00 L of ocean water having 3.0×10^{-4} ppm silver (Ag^+)

(a) _____

(b) _____

ANSWERS AND SOLUTIONS TO PROBLEM 14-8

(a) 97 mg   In 81 ppm sulfate ions, there are 81 g sulfate ions in 1,000,000 g of sample. Calculate the number of milligrams of sulfate ions in 1.20 L of a water sample as follows:

$$1.20 \; \cancel{L \; sample} \times \frac{1000 \; \cancel{mL \; sample}}{1 \; \cancel{L \; sample}} \times \frac{1.00 \; \cancel{g \; sample}}{1 \; \cancel{mL \; sample}}$$

$$\times \frac{81 \; \cancel{g \; SO_4^{2-}}}{1,000,000 \; \cancel{g \; sample}} \times \frac{1000 \; mg \; SO_4^{2-}}{1 \; \cancel{g \; SO_4^{2-}}} = 97 \; mg \; SO_4^{2-}$$

(b) 1.8×10^{-3} mg

In 3.0×10^{-4} ppm of silver, there is 3.0×10^{-4} g silver in 1,000,000 g sample. Calculate the number of milligrams of silver in 6.00 L of ocean water as follows:

$$6.00 \ \cancel{\text{L sample}} \times \frac{1000 \ \cancel{\text{mL sample}}}{1 \ \cancel{\text{L sample}}} \times \frac{1.00 \ \cancel{\text{g sample}}}{1 \ \cancel{\text{mL sample}}}$$

$$\times \frac{3.0 \times 10^{-4} \ \cancel{\text{g Ag}^+}}{1,000,000 \ \cancel{\text{g sample}}} \times \frac{1000 \ \text{mg Ag}^+}{1 \ \cancel{\text{g Ag}^+}} = 18 \times 10^{-4} \ \text{mg Ag}^+$$

$$= 1.8 \times 10^{-3} \ \text{mg Ag}^+$$

If your answers are correct, proceed to Problem 14-9. If your answers are incorrect, review Section 14-7 (Example 14-8 and Study Exercise 14-6) in your text.

Proceed to Problem 14-9.

PROBLEM 14-9

Calculate the molarity of each of the following aqueous solutions:

(a)   4.75 g of sodium chloride in $25\overline{0}$ mL solution

(b)   15.0 g of calcium chloride in $35\overline{0}$ mL of solution. Also calculate the molarity of the chloride (Cl^-) ions

(a) _____

(b) _____

214

ANSWERS AND SOLUTIONS TO PROBLEM 14-9

(a) 0.325 $\underline{M}$ The molar mass of NaCl is 58.5 g; calculate the molarity as follows:

$$\frac{4.75 \text{ g NaCl}}{250 \text{ mL solution}} \times \frac{1 \text{ mol NaCl}}{58.5 \text{ g NaCl}} \times \frac{1000 \text{ mL solution}}{1 \text{ L solution}}$$

$$= \frac{0.325 \text{ mol NaCl}}{1 \text{ L solution}} = 0.325 \text{ } \underline{M}$$

(b) 0.386 $\underline{M}$ The molar mass of $CaCl_2$ is 111.1 g; calculate the
     0.772 $\underline{M}$ molarity as follows:

$$\frac{15.0 \text{ g CaCl}_2}{350 \text{ mL solution}} \times \frac{1 \text{ mol CaCl}_2}{111.1 \text{ g CaCl}_2} \times \frac{1000 \text{ mL solution}}{1 \text{ L solution}}$$

$$= \frac{0.386 \text{ mol NaCl}}{1 \text{ L solution}} = 0.386 \text{ } \underline{M}$$

One mole calcium chloride will form one mole of calcium ions and $\underline{two}$ moles of chloride ions according to the following balanced equation:

$CaCl_2(\underline{aq})$ ---> $Ca^{2+}(\underline{aq})$ + 2 $Cl^-(\underline{aq})$

In a 0.386 $\underline{M}$ calcium chloride solution there is 0.386 mol of calcium chloride per liter of solution. Hence, 0.386 mol of $CaCl_2$ will form 0.386 mol of Ca^{2+} ions and 0.772 mol Cl^- ions.

$$0.386 \text{ mol CaCl}_2 \times \frac{2 \text{ mol Cl}^-}{1 \text{ mol CaCl}_2} = 0.772 \text{ mol Cl}^-$$

The molarity of the chloride ion will be 0.772 $\underline{M}$.

If your answers are correct, proceed to Problem 14-10. If your answers are incorrect, review Section 14-8 (Example 14-9 and Study Exercise 14-7) in your text.

Proceed to Problem 14-10.

PROBLEM 14-10

(1) Calculate the number of grams of solute necessary to prepare the following aqueous solutions. (2) Explain how each solution would be prepared.

(a) $50\overline{0}$ mL of a 0.750 $\underline{M}$ sodium carbonate (Na_2CO_3) solution

(b) $35\overline{0}$ mL of a 0.100 $\underline{M}$ calcium chloride ($CaCl_2$) solution

(a) _____

(b) _____

ANSWERS AND SOLUTIONS TO PROBLEM 14-10

(a)   39.8 g   (1) The molar mass of Na_2CO_3 is 106.0 g. In a 0.750 $\underline{M}$ Na_2CO_3 solution, there is 0.750 mol Na_2CO_3 per 1.00 L of solution. Calculate the number of grams of Na_2CO_3 necessary for preparing 500 mL of a 0.750 $\underline{M}$ solution as follows:

500 ~~mL solution~~ x $\dfrac{1 \text{ ~~L solution~~}}{1000 \text{ ~~mL solution~~}}$ x $\dfrac{0.750 \text{ ~~mol Na}_2\text{CO}_3~~}{1 \text{ ~~L solution~~}}$

x $\dfrac{106.0 \text{ g Na}_2\text{CO}_3}{1 \text{ ~~mol Na}_2\text{CO}_3~~}$ = 39.8 g Na_2CO_3

(2) The sodium carbonate (39.8 g) is dissolved in sufficient water to make the total volume of the solution equal to 500 mL.

(b) 3.89 g (1) The molar mass of $CaCl_2$ is 111.1 g. In a 0.100 M $CaCl_2$ solution, there is 0.100 mol $CaCl_2$ per 1.00 L of solution. Calculate the number of grams of $CaCl_2$ necessary for preparing 350 mL of a 0.100 M solution as follows:

$$350 \text{ mL solution} \times \frac{1 \text{ L solution}}{1000 \text{ mL solution}} \times \frac{0.100 \text{ mol } CaCl_2}{1 \text{ L solution}}$$

$$\times \frac{111.1 \text{ g } CaCl_2}{1 \text{ mol } CaCl_2} = 3.89 \text{ g } CaCl_2$$

(2) The calcium chloride (3.89 g) is dissolved in sufficient water to make the total volume of the solution equal to 350 mL.

If your answers are correct, proceed to Problem 14-11. If your answers are incorrect, review Section 14-8 (Example 14-10 and Study Exercise 14-8) in your text.

Proceed to Problem 14-11.

PROBLEM 14-11

Calculate the number of milliliters of solution required to provide the following:

(a) 4.00 g of sodium chloride (NaCl) from a 0.120 M solution

(b) 1.75 g of sodium chromate (Na_2CrO_4) from a 1.20 M solution

(a) _____

(b) _____

217

ANSWERS AND SOLUTIONS TO PROBLEM 14-11

(a) $57\overline{0}$ mL   The molar mass of NaCl is 58.5 g. In a 0.120 M
NaCl solution, there is 0.120 mol NaCl per 1.00 L
of solution.   Calculate the number of milliliters
of 0.120 M solution necessary to provide 4.00 g of
NaCl as follows:

4.00 g̶ ̶N̶a̶C̶l̶ x $\dfrac{1 \text{ m̶o̶l̶ ̶N̶a̶C̶l̶}}{58.5 \text{ g̶ ̶N̶a̶C̶l̶}}$ x $\dfrac{1 \text{ L̶ ̶s̶o̶l̶u̶t̶i̶o̶n̶}}{0.120 \text{ m̶o̶l̶ ̶N̶a̶C̶l̶}}$

x $\dfrac{1000 \text{ mL solution}}{1 \text{ L̶ ̶s̶o̶l̶u̶t̶i̶o̶n̶}}$ = $57\overline{0}$ mL solution

(b) 9.00 mL The molar mass of Na_2CrO_4 is 162.0 g. In a 1.20 M

Na_2CrO_4 solution, there are 1.20 mol Na_2CrO_4 per

1.00 L of solution. Calculate the number of milli-

liters of 1.20 M solution necessary to provide

1.75 g Na_2CrO_4 as follows:

1.75 g̶ ̶N̶a̶₂̶C̶r̶O̶₄̶ x $\dfrac{1 \text{ m̶o̶l̶ ̶N̶a̶₂̶C̶r̶O̶₄̶}}{162.0 \text{ g̶ ̶N̶a̶₂̶C̶r̶O̶₄̶}}$ x $\dfrac{1 \text{ L̶ ̶s̶o̶l̶u̶t̶i̶o̶n̶}}{1.20 \text{ m̶o̶l̶ ̶N̶a̶₂̶C̶r̶O̶₄̶}}$

x $\dfrac{1000 \text{ mL solution}}{1 \text{ L̶ ̶s̶o̶l̶u̶t̶i̶o̶n̶}}$ = 9.00 mL solution

If your answers are correct, proceed to Problem 14-12. If your
answers are incorrect, review Section 14-8 (Example 14-11) in
your text.

Proceed to Problem 14-12

PROBLEM 14-12

Calculate the number of milliliters of 16.2 M concentrated
nitric acid stock solution needed to prepare the following
diluted nitric acid solutions:

(a) $75\overline{0}$ mL of a 6.00 M nitric acid solution

(b) $30\overline{0}$ mL of a 5.00 M nitric acid solution

ANSWERS AND SOLUTIONS TO PROBLEM 14-12

(a) 278 mL (1) Calculate the moles of nitric acid in the diluted solution, using 6.00 M (6.00 mol HNO_3 /1 L
solution and $75\overline{0}$ mL of solution:

$75\overline{0}$ m̶L̶ ̶s̶o̶l̶u̶t̶i̶o̶n̶ x $\dfrac{1 \text{ L̶ ̶s̶o̶l̶u̶t̶i̶o̶n̶}}{1000 \text{ m̶L̶ ̶s̶o̶l̶u̶t̶i̶o̶n̶}}$ x $\dfrac{6.00 \text{ mol } HNO_3}{1 \text{ L̶ ̶s̶o̶l̶u̶t̶i̶o̶n̶}}$ = 4.50 mol HNO_3

218

(2) The 4.50 mol HNO_3 needed is obtained from the stock solution consisting of 16.2 M (16.2 mol HNO_3/1 L solution). The number of milliliters of the stock solution is calculated as

4.50 mol HNO_3 x $\dfrac{1 \text{ L solution}}{16.2 \text{ mol } HNO_3}$ x $\dfrac{1000 \text{ mL solution}}{1 \text{ L solution}}$ = 278 mL

(b) 92.6 mL (1) Calculate the moles of nitric acid in the diluted solution, using 5.00 M (5.00 mol HNO_3/ 1 L solution and 300 mL of solution:

300 mL solution x $\dfrac{1 \text{ L solution}}{1000 \text{ mL solution}}$ x $\dfrac{5.00 \text{ mol } HNO_3}{1 \text{ L solution}}$ = 1.50 mol HNO_3

(2) The 1.50 mol HNO_3 needed is obtained from the stock solution consisting of 16.2 M (16.2 mol HNO_3/1 L solution). The number of milliliters of the stock solution is calculated as

1.50 mol HNO_3 x $\dfrac{1 \text{ L solution}}{16.2 \text{ mol } HNO_3}$ x $\dfrac{1000 \text{ mL solution}}{1 \text{ L solution}}$ = 92.6 mL

If your answers are correct, proceed to Problem 14-13. If your answers are incorrect, review Section 14-8, dilution of molar solutions (Examples 14-12 and 14-13 and Study Exercise 14-9) in your text.

Proceed to Problem 14-13.

PROBLEM 14-13

Calculate the normality of each of the following aqueous solutions:

(a) 7.85 g nitric acid (HNO_3) in 400 mL of solution

(b) 43.0 g sulfuric acid (H_2SO_4) in 500 mL of solution in

reactions that replace both hydrogen ions.

(a) _____

(b) _____

(a) 0.312 $\underline{N}$ The molar mass of HNO_3 is 63.0 g, and because 1 mol of hydrogen ions is used per mole of the acid, one equivalent of HNO_3 is equal to 63.0 g (63.0 g/1).

$$\frac{7.85 \text{ g } HNO_3}{400 \text{ mL solution}} \times \frac{1 \text{ eq } HNO_3}{63.0 \text{ g } HNO_3} \times \frac{1000 \text{ mL solution}}{1 \text{ L solution}}$$

$$= \frac{0.312 \text{ eq } HNO_3}{1.00 \text{ L solution}} = 0.312 \underline{N}$$

(b) 1.76 $\underline{N}$ The molar mass of H_2SO_4 is 98.1 g, and because 2 mol of hydrogen ions are used per mole of the acid, one equivalent of H_2SO_4 is equal to 49.0 g (98.1 g/2).

$$\frac{43.0 \text{ g } H_2SO_4}{500 \text{ mL solution}} \times \frac{1 \text{ eq } H_2SO_4}{49.0 \text{ g } H_2SO_4} \times \frac{1000 \text{ mL solution}}{1 \text{ L solution}}$$

$$= \frac{1.76 \text{ eq } H_2SO_4}{1.00 \text{ L solution}} = 1.76 \underline{N}$$

If your answers are correct, proceed to Problem 14-14. If your answers are incorrect, review Section 14-9 (Example 14-14 and Study Exercise 14-10) in your text.

Proceed to Problem 14-14.

PROBLEM 14-14

Calculate the number of grams of solute necessary to prepare the following aqueous solutions:

(a) 45.0 mL of a 2.45 $\underline{N}$ sodium hydroxide solution

(b) $35\overline{0}$ mL of a 0.125 $\underline{N}$ sulfuric acid solution in reactions that replace both hydrogen ions

(a) _____

(b) _____

ANSWERS AND SOLUTIONS TO PROBLEM 14-14

(a) 4.41 g   The molar mass of NaOH is 40.0 g, and because 1 mol of hydroxide ions is used per mole of the base, one equivalent of NaOH is equal to 40.0 g (40.0 g/1). In a 2.45 $\underline{N}$ NaOH solution, there would be 2.45 eq of NaOH in 1.00 L of solution. Therefore, the number of grams of NaOH necessary to prepare 45.0 mL of a 2.45 $\underline{N}$ NaOH solution would be:

$$45.0 \; \overline{mL \; solution} \; \times \; \frac{1 \; \overline{L \; solution}}{1000 \; \overline{mL \; solution}} \; \times \; \frac{2.45 \; \overline{eq \; NaOH}}{1 \; \overline{L \; solution}}$$

$$\times \; \frac{40.0 \; g \; NaOH}{1 \; \overline{eq \; NaOH}} = 4.41 \; g \; NaOH$$

(b) 2.14 g   The molar mass of H_2SO_4 is 98.1 g; because 2 mol of hydrogen ions are used per mole of the acid, one equivalent of H_2SO_4 is equal to 49.0 g (98.1 g/2). In a 0.125 $\underline{N}$ H_2SO_4 solution, there would be 0.125 eq of H_2SO_4 in 1.00 L of solution. Therefore, the number of grams of H_2SO_4 necessary to prepare 35$\overline{0}$ mL of a 0.125 $\underline{N}$ H_2SO_4 solution would be:

$$35\overline{0} \; \overline{mL \; solution} \; \times \; \frac{1 \; \overline{L \; solution}}{1000 \; \overline{mL \; solution}} \; \times \; \frac{0.125 \; \overline{eq \; H_2SO_4}}{1 \; \overline{L \; solution}}$$

$$\times \; \frac{49.0 \; g \; H_2SO_4}{1 \; \overline{eq \; H_2SO_4}} = 2.14 \; g \; H_2SO_4$$

If your answers are correct, proceed to Problem 14-15. If your answers are incorrect, review Section 14-9 (Example 14-15) in your text.

Proceed to Problem 14-15.

PROBLEM 14-15

Calculate the number of milliliters of an aqueous solution required to provide the following:

(a) $11\overline{0}$ g of sulfuric acid (H_2SO_4) from a 2.50 <u>N</u> solution in

reactions that replace both hydrogen ions

(b) $12\overline{0}$ g of phosphoric acid (H_3PO_4) from a 3.00 <u>N</u> solution

in reactions that replace all three hydrogen ions.

(a) _____

(b) _____

ANSWERS AND SOLUTIONS TO PROBLEM 14-15

(a) 898 mL   A 2.50 <u>N</u> sulfuric acid solution would contain 2.50
             eq/L of solution. Because 2 mol of hydrogen ions
             are used per mole of the acid, one equivalent of
             H_2SO_4 is equal to 49.0 g (98.1 g/2). Calculate the

             number of milliliters as follows:

$$11\overline{0} \; \cancel{g \; H_2SO_4} \times \frac{1 \; \cancel{eq \; H_2SO_4}}{49.0 \; \cancel{g \; H_2SO_4}} \times \frac{1 \; \cancel{L \; solution}}{2.50 \; \cancel{eq \; H_2SO_4}}$$

$$\times \frac{1000 \; mL \; solution}{1 \; \cancel{L \; solution}} = 898 \; mL \; solution$$

(b) 1,220 mL   A 3.00 N phosphoric acid solution would contain
               3.00 eq/L of solution. Because 3 mol of hydrogen
               ions are used per mole of the acid, one equiva-
               lent of H_3PO_4 is equal to 32.7 g (98.0 g/3). Cal-
               culate the number of milliliters as follows:

$$120 \text{ g } H_3PO_4 \times \frac{1 \text{ eq } H_3PO_4}{32.7 \text{ g } H_3PO_4} \times \frac{1 \text{ L solution}}{3.00 \text{ eq } H_3PO_4}$$

$$\times \frac{1000 \text{ mL solution}}{1 \text{ L solution}} = 1{,}220 \text{ mL}$$ solution (to three significant digits)

If your answers are correct, proceed to Problem 14-16. If your answers are incorrect, review Section 14-9 (Example 14-16 and Study Exercise 14-11) in your text.

Proceed to Problem 14-16.

PROBLEM 14-16

Calculate the molarity of the following solutions:

(a) 6.00 $\underline{N}$ sulfuric acid solution in reactions that replace both hydrogen ions

(b) 1.50 $\underline{N}$ phosphoric acid solution in reactions that replace all three hydrogen ions.

ANSWERS AND SOLUTIONS TO PROBLEM 14-16

(a) 3.00 $\underline{M}$ The formula of sulfuric acid is H_2SO_4. Since both hydrogen ions are replaced there are 2 eq H_2SO_4/ 1 mol H_2SO_4. Solving for moles per liter, using 6.00 eq H_2SO_4/1 L solution as follows:

$$\frac{6.00 \text{ eq } H_2SO_4}{1 \text{ L solution}} \times \frac{1 \text{ mol } H_2SO_4}{2 \text{ eq } H_2SO_4} = \frac{3.00 \text{ mol } H_2SO_4}{1 \text{ L solution}} = 3.00 \underline{M}$$

(b) 0.500 $\underline{M}$ The formula of phosphoric acid is H_3PO_4. Since all three hydrogen ions are replaced there are 3 eq H_3PO_4/ 1 mol H_3PO_4. Solving for moles per liter, using 1.50 eq H_2SO_4/1 L solution as follows:

$$\frac{1.50 \text{ eq } H_3PO_4}{1 \text{ L solution}} \times \frac{1 \text{ mol } H_3PO_4}{3 \text{ eq } H_3PO_4} = \frac{0.500 \text{ mol } H_3PO_4}{1 \text{ L solution}} = 0.500 \underline{M}$$

If your answers are correct, proceed to Problem 14-17. If your answers are incorrect, review 14-9, relation of molarity to normality (Example 14-17 and Study Exercise 14-12) in your text.

Proceed to Problem 14-17.

PROBLEM 14-17

Calculate the molality of each of the following solutions:

(a)   125 g of sulfuric acid (H_2SO_4) in $75\overline{0}$ g of water

(b)   20.0 g of glucose ($C_6H_{12}O_6$) in 75.0 g of water

(a) _____

(b) _____

ANSWERS AND SOLUTIONS TO PROBLEM 14-17

(a)   1.70 $\underline{m}$ The molar mass of H_2SO_4 is 98.1 g.   Calculate the molality as follows:

$$\frac{125 \text{ g } H_2SO_4}{750 \text{ g } H_2O} \times \frac{1 \text{ mol } H_2SO_4}{98.1 \text{ g } H_2SO_4} \times \frac{1000 \text{ g } H_2O}{1 \text{ kg } H_2O}$$

$$= \frac{1.70 \text{ mol } H_2SO_4}{1 \text{ kg } H_2O} = 1.70 \underline{m}$$

(b)   1.48 $\underline{m}$ The molar mass of glucose ($C_6H_{12}O_6$) is $18\overline{0}$ g. Calculate the molality as follows:

$$\frac{20.0 \text{ g } C_6H_{12}O_6}{75.0 \text{ g } H_2O} \times \frac{1 \text{ mol } C_6H_{12}O_6}{180 \text{ g } C_6H_{12}O_6} \times \frac{1000 \text{ g } H_2O}{1 \text{ kg } H_2O}$$

$$\frac{1.48 \text{ mol } C_6H_{12}O_6}{1 \text{ kg } H_2O} = 1.48 \underline{m}$$

If your answers are correct, proceed to Problem 14-1. If your answers are incorrect, review Section 14-10 (Example 14-18 and Study Exercise 14-13) your text.

Proceed to Problem 14-18.

PROBLEM 14-18

Calculate the number of grams of solute necessary to prepare the following aqueous solutions:

(a) $40\overline{0}$ g of a 1.20 $\underline{m}$ solution of sulfuric acid (H_2SO_4)

(b) $75\overline{0}$ g of a 4.00 $\underline{m}$ solution of sodium nitrate ($NaNO_3$)

(a) _____

(b) _____

ANSWERS AND SOLUTIONS TO PROBLEM 14-18

(a) 42.2 g The molar mass of H_2SO_4 is 98.1 g. A 1.20 $\underline{m}$ sulfuric acid solution would contain 1.20 mol

$$(1.20 \, \cancel{mol} \times \frac{98.1 \text{ g}}{1 \, \cancel{mol}} = 118 \text{ g})$$ of sulfuric acid in

1.00 kg (1000 g) of water. The total mass of the solution would be 1118 g (118 g sulfuric acid + $10\overline{00}$ g H_2O). Calculate the mass of sulfuric acid needed for $40\overline{0}$ g of a 1.20 $\underline{m}$ solution as follows:

$$40\overline{0} \, \cancel{\text{g solution}} \times \frac{118 \text{ g } H_2SO_4}{1118 \, \cancel{\text{g solution}}} = 42.2 \text{ g } H_2SO_4$$

(b) $19\overline{0}$ g The molar mass of $NaNO_3$ is 85.0 g. A 4.00 $\underline{m}$ sodium nitrate solution would contain 4.00 mol

$$(4.00 \, \cancel{mol} \times \frac{85.0 \text{ g}}{1 \, \cancel{mol}} = 34\overline{0} \text{ g})$$ of sodium

225

nitrate in 1.00 kg ($\overline{1000}$ g) of water. The total mass of the solution would be 134$\overline{0}$ g (34$\overline{0}$ g sodium nitrate + 1$\overline{000}$ g water). Calculate the mass of sodium nitrate needed for 75$\overline{0}$ g of a 4.00 $\underline{m}$ solution as follows:

$$75\overline{0} \text{ g solution} \times \frac{34\overline{0} \text{ g NaNO}_3}{134\overline{0} \text{ g solution}} = 19\overline{0} \text{ g NaNO}_3$$

If your answers are correct, proceed to Problem 14-19. If your answers are incorrect, review Section 14-10 (Example 14-19 and Study Exercise 14-14) in your text.

Proceed to Problem 14-19.

PROBLEM 14-19

Calculate the number of grams of water that must be added to:

(a) 30.0 g sodium chloride in the preparation of a 1.20 $\underline{m}$ solution

(b) 25.0 g glucose ($C_6H_{12}O_6$) in the preparation of a 3.50 $\underline{m}$ solution

(a) _____

(b) _____

(a)   427 g   The question asks for the number of grams of water; therefore you must use the inverse factor for molality. The molar mass of sodium chloride is 58.5 g. Calculate the number of grams of water as follows:

$$30.0 \text{ g NaCl} \times \frac{1 \text{ mol NaCl}}{58.5 \text{ g NaCl}} \times \frac{1 \text{ kg H}_2\text{O}}{1.20 \text{ mol NaCl}}$$

$$\frac{1000 \text{ g H}_2\text{O}}{1 \text{ kg H}_2\text{O}} = 427 \text{ g H}_2\text{O}$$

(b)   39.7 g   The molar mass of glucose ($C_6H_{12}O_6$) is $18\overline{0}$ g. Calculate the number of grams of water as follows:

$$25.0 \text{ g C}_6\text{H}_{12}\text{O}_6 \times \frac{1 \text{ mol C}_6\text{H}_{12}\text{O}_6}{180 \text{ g C}_6\text{H}_{12}\text{O}_6} \times \frac{1 \text{ kg H}_2\text{O}}{3.50 \text{ mol C}_6\text{H}_{12}\text{O}_6}$$

$$\times \frac{1000 \text{ g H}_2\text{O}}{1 \text{ kg H}_2\text{O}} = 39.7 \text{ g H}_2\text{O}$$

If your answers are correct, proceed to Problem 14-20. If your answers are incorrect, review Section 14-10 (Example 14-20 and Study Exercise 14-15) in your text.

Proceed to Problem 14-20.

PROBLEM 14-20

Calculate the boiling point (1 atm) and the freezing point of the following solutions (see Table 14-3 in your text for additional data):

(a)   a 1.85 molal aqueous sugar ($C_{12}H_{22}O_{11}$) solution

(b)   a sugar solution containing 9.25 g of sugar ($C_{12}H_{22}O_{11}$) in 75.0 g water

(a) _____

ANSWERS AND SOLUTIONS TO PROBLEM 14-20

(a) 100.96°C Calculate the boiling-point elevation from the equation:

-3.44°C $\triangle T_b$ = molality (m) x K_b = 1.85 m x $\dfrac{0.52°C}{1 \, m}$

$$= 0.96°C$$

(The value of K_b for water is in Table 14-3 of your text.)

The boiling point of the solution (1 atm) is 100.96°C (100.00°C + 0.96°C = 100.96°C). Calculate the freezing-point depression from the equation:

$\triangle T_f$ = molality (m) x K_f = 1.85 m x $\dfrac{1.86°C}{1 \, m}$ = 3.44°C

(The value of K_f for water is in Table 14-3 of

your text.) The freezing point of the solution is -3.44°C (0.00°C - 3.44°C = -3.44°C).

(b) 100.19°C The molar mass of sugar ($C_{12}H_{22}O_{11}$) is 342 g.

-0.67°C Calculate the molality of the solution as follows:

$$\dfrac{9.25 \text{ g } C_{12}H_{22}O_{11}}{75.0 \text{ g } H_2O} \times \dfrac{1 \text{ mol } C_{12}H_{22}O_{11}}{342 \text{ g } C_{12}H_{22}O_{11}} \times \dfrac{1000 \text{ g } H_2O}{1 \text{ kg } H_2O}$$

$$\dfrac{0.361 \text{ mol } C_{12}H_{22}O_{11}}{1 \text{ kg } H_2O} = 0.361 \text{ m}$$

Calculate the boiling-point elevation as follows:

$\triangle T_b$ = 0.361 m x $\dfrac{0.52°C}{1 \, m}$ = 0.19°C

Therefore, the boiling point of the solution (1 atm) is 100.19°C (100.00°C + 0.19°C = 100.19°C). Calculate the freezing-point depression as follows:

$$\triangle T_f = 0.361 \underline{m} \times \frac{1.86^{O}C}{1 \underline{m}} = 0.67^{O}C$$

Therefore, the freezing point of the solution is $-0.67^{O}C$ $(0.00^{O}C - 0.67^{O}C = -0.67^{O}C)$.

If your answers are correct, proceed to Problem 14-21. If your answers are incorrect, review Section 14-11, boiling-point elevation and freezing-point depression (Examples 14-21 and 14-22 and Study Exercises 14-16 and 14-17) in your text.

Proceed to Problem 14-21.

PROBLEM 14-21

Calculate the molecular mass and molar mass of a nonvolatile-nonionized unknown given the following data and additional information given in Table 14-3 of your text:

(a)   6.25 g of the unknown was dissolved in 125 g of water with the resulting solution having a freezing point of $-1.46^{O}C$

(b)   3.50 g of the unknown was dissolved in 100 g of benzene with the resulting solution having a freezing point of $4.50^{O}C$

(a) _____

(b) _____

(a) 63.7 amu, The freezing point of the solution is $-1.46^{\circ}C$.
    63.7 g The $\triangle T_f$ is $1.46^{\circ}C$ [freezing point of water (see Table 14-3 in your text) - freezing point of the solution = $0.00^{\circ}C - (-1.46^{\circ}C) = 1.46^{\circ}C$]. Calculate the molality of the solution as follows:

$$\text{molality (}\underline{m}\text{)} = \frac{\triangle T_f}{K_f} = 1.46^{\circ}\text{C} \times \frac{1\ m}{1.86^{\circ}\text{C}} = 0.785\ \underline{m}$$

$$= 0.785\ \frac{\text{mol of unknown}}{1\ \text{kg of water}}$$

(The value of K_f for water is in Table 14-3 of your text.) Calculate the molecular mass and molar mass for the unknown, solving for grams per mole (g/mol), as follows:

$$\frac{6.25\ g\ \text{unknown}}{125\ g\ \text{water}} \times \frac{1000\ g\ H_2O}{1\ kg\ H_2O} \times \frac{1\ kg\ H_2O}{0.785\ \text{mol}}$$

$$= 63.7\ \text{g/mol, } 63.7\ \text{amu, } 63.7\ g$$

(b) 179 amu, The freezing point of the solution is $4.50^{\circ}C$.
    179 g The T_f is $1.00^{\circ}C$ [freezing point of benzene (see Table 14-3 in your text) - freezing point of the solution = $5.50^{\circ}C - 4.50^{\circ}C = 1.00^{\circ}C$]. Calculate the molality of the solution as follows:

$$\text{molality} = \frac{\triangle T_f}{K_f} = 1.00^{\circ}\text{C} \times \frac{1\ m}{5.12^{\circ}\text{C}} = 0.195\ \underline{m}$$

$$= 0.195\ \frac{\text{mol of unknown}}{1\ \text{kg of benzene}}$$

(The value of K_f for benzene is in Table 14-3 of your text.) Calculate the molecular mass and molar mass of the unknown, solving for grams per mole (g/mol), as follows:

$$\frac{3.50\ g\ \text{unknown}}{100\ g\ \text{benzene}} \times \frac{1000\ g\ \text{benzene}}{1\ kg\ \text{benzene}} \times \frac{1\ kg\ \text{benzene}}{0.195\ \text{mol}}$$

$$= 179\ \text{g/mol, } 179\ \text{amu, } 179\ g$$

If your answers are correct, you have completed this chapter. If your answers are incorrect, review Section 14-11, determination of molecular mass (Example 14-23 and Study Exercise 14-18) in your text.

These problems conclude the chapter on solutions.

Now take the sample quizzes to see if you have mastered the material in this chapter.

SAMPLE QUIZZES

The quizzes divide this chapter into two parts. Quiz 14A (Sections 14-1 to 14-8) and Quiz 14B (Sections 14-9 to 14-12).

Quiz #14A (Sections 14-1 to 14-8)

Element	Atomic Mass (amu)
Na	23.0
Cl	35.5

1. Calculate the percent by mass of sodium chloride in a solution containing 11.4 g sodium chloride in 75.0 g of water.

2. Calculate the number of grams of sodium chloride that must be added to 60.0 g of water to prepare a 15.0 % by mass sodium chloride solution.

3. Calculate the parts per million (ppm) sodium ions containing 220 mg of sodium ions in 750 mL of a water sample.

4. Calculate the molarity of a solution of sodium chloride containing 39.7 g sodium chloride in 650 mL of solution.

5. Calculate the number of grams of sodium chloride needed to prepare 525 mL of a 0.300 $\underline{M}$ sodium chloride solution.

Solutions and Answers for Quiz #14A

1. Total mass of solution = 11.4 g NaCl + 75.0 g water = 86.4 g solution

$$\frac{11.4 \text{ g NaCl}}{86.4 \text{ g solution}} \times 100 = 13.2\%$$

2. 100.0 g solution - 15.0 g NaCl = 85.0 g water

$$60.0 \text{ g } H_2O \times \frac{15.0 \text{ g NaCl}}{85.0 \text{ g } H_2O} = 10.6 \text{ g NaCl}$$

3. $$\frac{22\overline{0} \text{ mg Na}^+}{75\overline{0} \text{ mL water sample}} \times \frac{1 \text{ mL sample}}{1.00 \text{ g sample}} \times \frac{1 \text{ g sample}}{1000 \text{ mg sample}}$$

$$\times 1,000,000 = 293 \text{ ppm}$$

4. $$\frac{39.7 \text{ g NaCl}}{650 \text{ mL solution}} \times \frac{1 \text{ mol NaCl}}{58.5 \text{ g NaCl}} \times \frac{1000 \text{ mL solution}}{1 \text{ L solution}}$$

$$= \frac{1.04 \text{ mol NaCl}}{1 \text{ L solution}} = 1.04 \underline{M}$$

5. $$525 \text{ mL solution} \times \frac{1 \text{ L solution}}{1000 \text{ mL solution}} \times \frac{0.300 \text{ mol NaCl}}{1 \text{ L solution}} \times$$

$$\frac{58.5 \text{ g NaCl}}{1 \text{ mol NaCl}} = 9.21 \text{ g NaCl}$$

Quiz # 14B (Sections 14-9 to 14-12)

Element	Atomic Mass (amu)
P	31.0
H	1.0
O	16.0
Na	23.0
Cl	35.5
S	32.1

1. Calculate the number of grams of phosphoric acid necessary to prepare 31̄0 mL of a 0.100 $\underline{N}$ phosphoric acid solution in reactions that replace all three hydrogen ions.

2. Calculate the normality of a 4.50 $\underline{M}$ sulfuric acid solution in reactions that replace both hydrogen ions.

3. Calculate the number of grams of sodium chloride necessary to prepare 265 g of a 0.325 $\underline{m}$ solution of sodium chloride.

4. Calculate the number of grams of water that must be added to 6.00 g of glucose ($C_6H_{12}O_6$) to prepare a 0.300 $\underline{m}$ glucose solution.

5. Calculate the molecular mass and molar mass of a non-volatile-nonionized unknown if a solution containing 7.80 g of the unknown in 15̄0 g of water has a freezing point of $-1.45^{\circ}C$. (Freezing point water = $0.00^{\circ}C$, K_f for water = $1.86^{\circ}C/\underline{m}$)

1. $31\overline{0}$ mL solution x $\dfrac{1\text{ L solution}}{1000\text{ mL solution}}$ x $\dfrac{0.100\text{ eq } H_3PO_4}{1\text{ L solution}}$

$= \dfrac{32.7\text{ g } H_3PO_4}{1\text{ eq } H_3PO_4} = 1.01\text{ g } H_3PO_4$

$$\begin{aligned}
3 \times 1.0 &= 3.0\text{ amu} \\
1 \times 31.0 &= 31.0\text{ amu} \\
4 \times 16.0 &= \underline{64.0\text{ amu}} \\
\text{FM for } H_3PO_4 &= 98.0\text{ amu}
\end{aligned}$$

1 eq H_3PO_4 = 98.0 g/3 = 32.7 g

2. $\dfrac{4.50\text{ mol } H_2SO_4}{1\text{ L solution}}$ x $\dfrac{2\text{ eq } H_2SO_4}{1\text{ mol } H_2SO_4} = \dfrac{9.00\text{ eq } H_2SO_4}{1\text{ L solution}} = 9.00\text{ }\underline{N}\text{ } H_2SO_4$

3. 0.325 mol NaCl x $\dfrac{58.5\text{ g NaCl}}{1\text{ mol NaCl}}$ = 19.0 g NaCl

(molar mass NaCl = 58.5 g)

$10\overline{00}$ g H_2O + 19.0 g NaCl = 1019 g solution

265 g solution x $\dfrac{19.0\text{ g NaCl}}{1019\text{ g solution}}$ = 4.94 g NaCl

4. The molar mass of glucose ($C_6H_{12}O_6$) is $18\overline{0}$ g.

$$\begin{aligned}
6 \times 12.0 &= 72.0\text{ amu} \\
12 \times 1.0 &= 12\text{ amu} \\
6 \times 16.0 &= \underline{96.0\text{ amu}} \\
\text{molecular mass} &= 18\overline{0}\text{ amu}
\end{aligned}$$

6.00 g $C_6H_{12}O_6$ x $\dfrac{1\text{ mol } C_6H_{12}O_6}{18\overline{0}\text{ g } C_6H_{12}O_6}$ x $\dfrac{1\text{ kg } H_2O}{0.300\text{ mol } C_2H_{12}O_6}$ x

$\dfrac{1000\text{ g } H_2O}{1\text{ kg } H_2O}$ = 111 g H_2O

5. $\triangle T_f = 0.00^{\circ}C - (-1.45^{\circ}C) = 1.45^{\circ}C$

molality ($\underline{m}$) $= \dfrac{\triangle T_f}{K_f} = 1.45^{\circ}C$ x $\dfrac{1\text{ }\underline{m}}{1.86^{\circ}C}$ = 0.780 $\underline{m}$

$= \dfrac{0.780\text{ mol of unknown}}{1\text{ kg of water}}$

$\dfrac{7.80\text{ g of unknown}}{15\overline{0}\text{ g } H_2O}$ x $\dfrac{1000\text{ g } H_2O}{1\text{ kg } H_2O}$ x $\dfrac{1\text{ kg } H_2O}{0.780\text{ mol}}$

= 66.7 g/mol, 66.7 amu, 66.7 g

CHAPTER 15

ACIDS, BASES, AND IONIC EQUATIONS

In Chapter 7 we discussed the nomenclature of acids and bases. In Chapter 9 the reaction of an acid and a base as an example of a neutralization reaction was considered. In this chapter acids and bases are discussed in more exact terms and their properties are examined. In Chapter 9 we considered equations written as complete (or molecular equations); in this chapter we will also consider these equations in ionic form - ionic equations.

SELECTED TOPICS

1. Two definitions of acids and bases that we will consider are: (1) Arrhenius definition and (2) Brønsted-Lowry definition.

    An <u>Arrhenius acid</u> is a substance that releases hydrogen ions (H^+) when dissolved in water. Generally the hydrogen ion is hydrated in aqueous solution as the hydronium ion (H_3O^+) according to the following equation:

$$H^+ + H_2O \colon\ \text{---}> \quad H_3O^+$$

    According to this definition, sulfuric acid (H_2SO_4), acetic acid ($HC_2H_3O_2$), and nitric acid (HNO_3) are some examples of Arrhenius acids. An <u>Arrhenius base</u> is a substance that releases hydroxide (OH^-) ions when dissolved in water. According to this definition, potassium hydroxide (KOH), sodium hydroxide (NaOH), magnesium hydroxide [$Mg(OH)_2$], calcium hydroxide [$Ca(OH)_2$], and aluminum hydroxide [$Al(OH)_3$] are examples of Arrhenius bases.

    A <u>Brønsted-Lowry</u> <u>acid</u> is any substance that donates a proton to some other substance. A <u>Brønsted-Lowry</u> base is any substance that accepts a proton from another substance. Any substance that is an Arrhenius acid or base will also be a Brønsted-Lowry acid or base. In addition, ions or uncharged molecules can act as Brønsted-Lowry acids or bases. Substances capable of acting as either a Brønsted-Lowry acid or base are amphoteric substances. Therefore, an <u>amphoteric substance</u>, is any substance that can act as an acid or base, depending on the nature of the solution. Examples of amphoteric substances are H_2O and HSO_4^-.

235

When an acid loses a proton, the species that is produced is called a conjugate base. For example when HNO_3 (nitric acid) loses a proton it forms NO_3^- (nitrate ion) which is the conjugate base. Therefore, a <u>conjugate base</u> is the species formed when a proton (H^+) is <u>removed</u> from an acid. When a base gains a proton, the species that is produced is called a conjugate acid. For example when OH^- (hydroxide ion), gains a proton it forms H_2O which is the conjugate acid. Therefore, a <u>conjugate acid</u> is the species formed when a proton (H^+) is added to a base. See Section 15-1 in your text.

2. Acids can be either strong or weak depending on their ability to donate a proton to another molecule or ion. Bases can be either strong or weak depending on their ability to give hydroxide ions. Table 15-2 in your text lists six strong acids and eight strong bases. All other acids and bases are considered to be weak. See Section 15-2 in your text.

3. Water is an amphoteric substance. It can react with itself to form hydronium or hydrogen ions and hydroxide ions. The concentrations of these ions in pure water at $25^{\circ}C$ are 1×10^{-7} mol/L of both hydrogen and hydroxide ions. A solution whose concentration of hydrogen ions is greater than 1×10^{-7} mol/L is considered acid and a solution whose concentration of hydrogen ions is less than 1×10^{-7} mol/L or hydroxide ions is greater than 1×10^{-7} mol/L is considered basic. When the concentration of both hydrogen ions and hydroxide ions is equal to 1×10^{-7} mol/L, the solution is considered neutral.

The formation of ions from compounds can occur in two different ways: (1) dissociation or (2) ionization. In <u>dissociation</u>, the ions in an ionic compound separate when acted upon by the solvent. In <u>ionization</u>, covalent compounds dissolve to produce ions in aqueous solution. See Section 15-3 in your text.

4. <u>Titration</u> is a procedure for determining the concentration of an acid or base by adding a base or an acid of known concentration. The point at which an acid or base (original substance) is exactly neutralized in a titration is the <u>equivalence point</u>. The equivalence point is usually determined by a change in color of an appropriately selected indicator. <u>Indicators</u> are substances that change color

in acid or base. Phenolphthalein, colorless in acid and red in base, is an often-used indicator. The point at which the indicator in an acid-base titration changes color is the end point. The end point is often the same as the the equivalence point, but not always.

You can determine the concentration of the unknown acid or base as follows:

(1) calculate moles of the unknown using the principles you learned in Chapter 10 on stoichiometry

(2) calculate the concentration of the unknown solution in moles per liter (molarity) from the moles of the unknown calculated in (1) and the volume of the solution given in the problem. If you need to find the concentration of the unknown solution in percent by mass, use the density of the solution.

In acid-base titrations, the number of equivalents of an acid is equal to the number of equivalents of a base. One

equivalent of an acid or base will supply 6.02×10^{23}

hydrogen ions (if an acid) or 6.02×10^{23} hydroxide ions (if a base). Therefore, one equivalent of any acid will combine exactly with one equivalent of any base. We have the following equation for this neutralization of any acid (A) by a base (B) or any base by an acid:

$$\text{equivalents A} = \text{equivalents B}$$

Using the definition of normality as equivalents/liter of solution, we can write the above equations as follows:

$$\text{eq A} = V_A \cdot N_A \ (eq_A/L_A)$$

$$\text{eq B} = V_B \cdot N_B \ (eq_B/L_B)$$

$$\text{eq A} = \text{eq B}$$

$$V_A \cdot N_A = V_B \cdot N_B$$

See Section 15-4 in your text.

5. pH is $- \log [H^+]$; pOH is $- \log [OH^-]$. The sum of the pH + pOH is equal to 14.00. A pH of 7 is neutral with 0 to 7 being acidic and 7 to 14 basic. You can calculate the pH or pOH of a solution given the hydrogen or hydroxide ion concentration in moles per liter using logarithms. Appendix II in your text will help you use your calculator to calculate logarithms of numbers. You can calculate the hydrogen ion or hydroxide ion concentration from the pH or

pOH of a solution using antilogarithms. See Appendix II in your text to help you use your calculator to calculate the antilogarithms. See Section 15-5 in your text.

6. Electrolytes are substances that in aqueous solutions can conduct an electric current. Nonelectrolytes are aqueous solutions that do not conduct an electric current. To determine whether a substance is an electrolyte or non-electrolyte, an aqueous solution of the substance is prepared and the solution is tested with two electrodes connected to a source of electric current (direct or alternating) with a standard light bulb in the circuit as shown in Figure 15-7 in your text. If the light bulb glows, the substance is an electrolyte; if it does not glow the substance is a nonelectrolyte. The reason for the conduction of an electric current by electrolytes in aqueous solution was explained by Arrhenius. He proposed that aqueous solutions of electrolytes contain ions while solutions of nonelectrolytes do not. Therefore, electrolytes produce ions in aqueous solution and conduct an electric current, while nonelectrolytes do not produce ions in aqueous solution and do not conduct an electric current. Salts, acids, and bases are electrolytes. Pure water, sugar (sucrose, $C_{12}H_{22}O_{11}$), ethyl alcohol (C_2H_6O), glycerine ($C_3H_8O_3$), and urea (CH_4N_2O) are nonelectrolytes. Electrolytes are further divided into strong and weak electrolytes. Strong electrolytes make the light bulb glow brightly; weak electrolytes give the light bulb a dull glow. Ionic salts such as sodium chloride (NaCl), strong acids such as sulfuric acid (H_2SO_4), hydrochloric acid (HCl), hydrobromic acid (HBr), hydriodic acid (HI), nitric acid (HNO_3), and perchloric acid ($HClO_4$), and strong bases such as group IA (1) hydroxides - sodium and potassium hydroxides (NaOH and KOH) - and other bases such as barium, strontium, and calcium hydroxides $Ba(OH)_2$, $Sr(OH)_2$, and $Ca(OH)_2$ are strong electrolytes. Weak electrolytes include weak acids and weak bases, such as acetic acid ($HC_2H_3O_2$), and ammonia water [$NH_3(aq)$]. Table 15-5 in your text lists strong and weak electrolytes and nonelectrolytes. To write ionic equations, you will need to recognize these compounds in Table 15-5 as strong or weak electrolytes, or nonelectrolytes. See Section 15-6 in your text.

7. Ionic equations express a chemical reaction involving compounds that exist mostly in ionic form in aqueous solution. In Section 15-7 of your text are guidelines for writing ionic equations. Study these guidelines. Notice in guideline 2 that the following formulas are written in **complete** form: **nonelectrolytes**, **weak acids** and **weak bases**, **solids** and **insoluble salts (precipitates)**, and **gases**. All strong acids and strong bases and soluble salts are written in ionic form. The net ionic equation shows only those ions that have actually undergone a chemical change. The ions that have not undergone a change and appear on both sides of the equation are crossed out. These ions are not

238

included in the net ionic equation. They are called "spectator ions" since they are present but are not involved in the actual reaction. They are included in the total ionic equation, but not in the net ionic equation. See Sections 15-7 and 15-8 in your text.

PROBLEM 15-1

Classify the following as an acid or base according to the Arrhenius theory:

(a) $HClO_3$ _____

(b) $Cd(OH)_2$ _____

(c) H_2SO_3 _____

(d) $Mg(OH)_2$ _____

ANSWERS TO PROBLEM 15-1

(a) acid

(b) base

(c) acid

(d) base

If your answers are correct, proceed to Problem 15-2. If your answers are incorrect, review Section 15-1, Arrhenius definition (Study Exercise 15-1) in your text.

Proceed to Problem 15-2

PROBLEM 15-2

Classify the following as an acid, base, or amphoteric substance according to the Brønsted-Lowry theory:

(a) H_2SO_3 _____

(b) HSO_3^- _____

(c) SO_3^{2-} _____

(d) H_2O _____

ANSWERS AND SOLUTIONS TO PROBLEM 15-2

(a) acid $H_2SO_3 \rightleftharpoons H^+ + HSO_3^-$ (acid)

(b)   amphoteric HSO_3^- $\xrightarrow{\text{----}}$ $\xleftarrow{\text{----}}$ $H^+ + SO_3^{2-}$ (acid)

$HSO_3^- + H^+$ $\xrightarrow{\text{--}}$ $\xleftarrow{\text{------}}$ H_2SO_3 (base)

Hence, HSO_3^- is an amphoteric substance.

(c)   base $SO_3^{2-} + H^+$ $\xrightarrow{\text{----}}$ $\xleftarrow{\text{----}}$ HSO_3^- (base)

(d)   amphoteric H_2O $\xrightarrow{\text{--}}$ $\xleftarrow{\text{----}}$ $H^+ + OH^-$ (acid)

$H_2O + H^+$ $\xrightarrow{\text{----}}$ H_3O^+ (base)

Hence, H_2O an is amphoteric substance.

If your answers are correct, proceed to Problem 15-3. If your answers are incorrect, review Section 15-1, Brønsted-Lowry definition (Study Exercise 15-2) in your text.

Proceed to Problem 15-3.

PROBLEM 15-3

From the following reaction equation, identify the conjugate acid and conjugate base:

$$HNO_3\,(\underline{aq}) \quad + \quad H_2O\,(\ell) \quad \text{----->} \quad H_3O^+\,(\underline{aq}) \quad + \quad NO_3^-\,(\underline{aq})$$

ANSWERS TO PROBLEM 15-3

Conjugate acid, H_3O^+; conjugate base, NO_3^-.

If your answers are correct, proceed to Problem 15-4. If your answers are incorrect, review Section 15-1, Brønsted-Lowry definition (Study Exercise 15-3) in your text.

Proceed to Problem 15-4.

PROBLEM 15-4

Classify each of the following substances as a strong acid, a weak acid, a strong base, or a weak base:

(a)   hydrobromic acid solution (HBr)

(b)   acetic acid solution ($HC_2H_3O_2$)

(c)   methylamine solution (CH_3-NH_2)

(d)   rubidium hydroxide solution (RbOH)

ANSWERS AND SOLUTIONS TO PROBLEM 15-3

(a) strong acid Hydrobromic acid solution (HBr) is listed in
                  Table 15-2 in your text as a strong acid.

(b) weak acid   Acetic acid solution ($HC_2H_3O_2$) is not listed
                  in Table 15-2 in your text. Therefore, it
                  is a weak acid.

(c) weak base   Methylamine solution (CH_3-NH_2) is not listed
                  in Table 15-2 in your text. Therefore, it
                  is a weak base. It is also a nitrogen-
                  containing compound similar to ammonia.

(d) strong base Rubidium hydroxide solution (RbOH) is listed
                  in Table 15-2 in your text as a strong base.

If your answers are correct, proceed to Problem 15-5. If your
answers are incorrect, review Section 15-2 (Study Exercise 15-
4) in your text.

Proceed to Problem 15-5

PROBLEM 15-5

Complete and balance the following chemical equations for the
following reactions:

(a) nitric acid + barium hydroxide ---->

(b) aluminum metal + nitric acid ---->

ANSWERS AND SOLUTIONS TO PROBLEM 15-5

(a) $2\ HNO_3(aq) + Ba(OH)_2(aq) \longrightarrow Ba(NO_3)_2(aq) + 2\ H_2O(\ell)$

    First, we must be able to write the correct formulas for
    nitric acid (HNO_3) and barium hydroxide [$Ba(OH)_2$]. Then,
    we can balance the equation.

(b) $2\ Al(s) + 6\ HNO_3(aq) \longrightarrow 2\ Al(NO_3)_3(aq) + 3\ H_2(g)$

    Again, we must write the correct formulas. Aluminum is
    above hydrogen in the electromotive or activity series, so
    aluminum will replace hydrogen to liberate hydrogen gas
    (H_2).

If your answers are correct, proceed to Problem 15-6. If your
answers are incorrect, review Section 15-4, reactions of acids
with metals, reactions of bases to form insoluble salts, and
neutralization reactions and titrations (Study Exercises 15-5,
15-6, and 15-7) in your text.

Proceed to Problem 15-6.

PROBLEM 15-6

If 37.6 mL of 0.320 $\underline{M}$ sodium hydroxide solution is necessary to neutralize 2.00 mL of a solution of hydrochloric acid to a phenolphthalein end point, calculate (a) the molarity and (b) the percent by mass (density of solution = 1.098 g/mL) of the hydrochloric acid solution.

(a) _____

(b) _____

ANSWERS AND SOLUTIONS TO PROBLEM 15-6

(a)   6.00 $\underline{M}$ The balanced equation is:

$$NaOH(\underline{aq}) + HCl(\underline{aq}) \longrightarrow NaCl(\underline{aq}) + H_2O(\underline{\ell})$$

First, we calculate the moles of hydrochloric acid neutralized with 37.6 mL of 0.320 $\underline{M}$ sodium hydroxide solution

$$37.6 \; \text{mL solution} \times \frac{1 \; \text{L}}{1000 \; \text{mL}} \times \frac{0.320 \; \text{mol NaOH}}{1 \; \text{L solution}}$$

$$\times \frac{1 \; \text{mol HCl}}{1 \; \text{mol NaOH}} = 0.0120 \; \text{mol HCl}$$

| Refers to the |
balanced equation

Next, we calculate the molarity of the hydrochloric acid solution from 2.00 mL of the hydrochloric acid solution that was used:

$$\frac{0.0120 \; \text{mol HCl}}{2.00 \; \text{mL solution}} \times \frac{1000 \; \text{mL solution}}{1 \; \text{L solution}} = \frac{6.00 \; \text{mol HCl}}{1 \; \text{L solution}}$$

$$= 6.00 \; \underline{M}$$

242

(b) 19.9 % To calculate the percent by mass of the hydrochloric acid solution, we must use the density of the solution (1.098 g/mL) and the molar mass of HCl (36.5 g).

$$\frac{6.00 \ \text{mol HCl}}{1 \ \text{L solution}} \times \frac{36.5 \ \text{g HCl}}{1 \ \text{mol HCl}} \times \frac{1 \ \text{L}}{1000 \ \text{mL}} \times$$

$$\frac{1 \ \text{mL solution}}{1.098 \ \text{g solution}} \times 100 = 19.9 \ \% \ \text{HCl}$$

If your answers are correct, proceed to Problem 15-7. If your answers are incorrect, review Section 15-4, neutralization reactions and titrations (Examples 15-1 and 15-2 and Study Exercise 15-8) in your text.

Proceed to Problem 15-7.

PROBLEM 15-7

If 0.425 g of pure sodium carbonate is titrated with 8.45 mL of hydrochloric acid solution to a methyl orange end point, calculate the molarity of the hydrochloric acid solution. (<u>Hint</u>: The methyl orange end point takes the carbonate ion to carbon dioxide.)

ANSWER AND SOLUTION TO PROBLEM 15-5

0.949 <u>M</u>   The balanced equation is:

$$Na_2CO_3 \ (\underline{s}) + 2 \ HCl \ (\underline{aq}) \longrightarrow 2 \ NaCl \ (\underline{aq}) + CO_2 \ (\underline{g}) + H_2O \ (\ell)$$

First, we calculate the moles of hydrochloric acid neutralized with 0.425 g pure sodium carbonate:

$$0.425 \ \text{g Na}_2\text{CO}_3 \times \frac{1 \ \text{mol Na}_2\text{CO}_3}{106.0 \ \text{g Na}_2\text{CO}_3} \times \frac{2 \ \text{mol HCl}}{1 \ \text{mol Na}_2\text{CO}_3}$$

= 0.00802 mol HCl | Refers to the |
                      | balanced equation |

(The molar mass of Na_2CO_3 is 106.0 g.)

243

Next, we calculate the molarity of the hydrochloric acid solution from 8.45 mL of the hydrochloric acid solution that was used.

$$\frac{0.00802 \text{ mol HCl}}{8.45 \text{ mL solution}} \times \frac{1000 \text{ mL}}{1 \text{ L}} = \frac{0.949 \text{ mol HCl}}{1 \text{ L solution}} = 0.949 \text{ M}$$

If your answer is correct, proceed to Problem 15-8. If your answer is incorrect, review Section 15-4, neutralization reactions and titrations (Examples 15-3) in your text.

Proceed to Problem 15-8.

PROBLEM 15-8

In the titration of 24.5 mL of sodium hydroxide solution of unknown concentration, 30.6 mL of 0.120 N sulfuric acid solution was required to neutralize the sodium hydroxide in reactions where both hydrogen ions of sulfuric acid react. Calculate the normality of the sodium hydroxide solution.

_____

ANSWER AND SOLUTION TO PROBLEM 15-8

0.150 N   We calculate the normality of the sodium hydroxide solution from the following equation:

$$V_A \cdot N_A = V_B \cdot N_B$$

with V_A = 0.0306 L H_2SO_4 solution,

$$N_A = \frac{0.120 \text{ eq } H_2SO_4}{L \ H_2SO_4 \text{ solution}} \quad ,$$

and V_B = 0.0245 L NaOH solution

$$0.0306 \text{ L } H_2SO_4 \text{ solution} \times \frac{0.120 \text{ eq } H_2SO_4}{L \ H_2SO_4 \text{ solution}}$$

$$= 0.0245 \text{ L NaOH solution} \times N_B$$

$$N_B = \frac{(0.0306 \text{ L})(0.120 \text{ eq/L})}{(0.0245 \text{ L})} = \frac{0.150 \text{ eq NaOH}}{L \text{ NaOH solution}} = 0.150 \text{ N}$$

244

If your answer is correct, proceed to Problem 15-9. If your answer is incorrect, review Section 15-4, neutralization reactions and titrations (Example 15-4 and Study Exercise 15-9) in your text.

Proceed to Problem 15-9.

PROBLEM 15-9

Calculate the pH and pOH of the following solutions:

(a)   hydrogen ion concentration is 2.2×10^{-8} mol/L

(b)   hydrogen ion concentration is 2.5×10^{-6} mol/L

(a) _____

(b) _____

ANSWERS AND SOLUTIONS TO PROBLEM 15-9

(a)   pH = 7.66   pH = $- \log[2.2 \times 10^{-8}]$

    pOH = 6.34   Use your calculator to determine the logarithm
of 2.2×10^{-8} as -7.6576 (see Appendix II in your text).

        pH = -[-7.6576] = 7.6576, 7.**66**
        pOH = 14.00 - 7.66 = 6.34

        (The logarithm of a number with two significant
digits will have **two** digits to the **right** of
the decimal point.)

(b)   pH = 5.60   pH = $- \log[2.5 \times 10^{-6}]$

    pOH = 8.40   Use your calculator to determine the logarithm
of 2.5×10^{-6} as -5.6021 (see Appendix II in your text).

        pH = -[-5.6021] = 5.6021, 5.**60**
        pOH = 14.00 - 5.60 = 8.40

        (The logarithm of a number with two significant
digits will have **two** digits to the **right** of
the decimal point.)

If your answers are correct, proceed to Problem 15-10. If your answers are incorrect, review Section 15-5 (Examples 15-5 and 15-6 and Study Exercise 15-10) in your text.

Proceed to Problem 15-10.

PROBLEM 15-10

Calculate the hydrogen ion concentration in moles per liter for each of the following solutions:

(a) a solution whose pH is 4.60

(b) a solution whose pH is 8.40

(a) _____

(b) _____

ANSWERS AND SOLUTIONS TO PROBLEM 15-10

(a) 2.5×10^{-5} mol/L
Because the pH of the solution is 4.60, $\log[H^+] = -4.\underline{60}$, we write the $[H^+] = 10^{-4.60}$.

Use your calculator to determine the antilogarithm of -4.60 as 2.51×10^{-5} (See Appendix II in your text).

$[H^+] = 2.51 \times 10^{-5}$ mol/L = **2.5** $\times 10^{-5}$ mol/L

(The antilogarithm of a number with <u>two</u> digits to the <u>right</u> of the decimal point will have **two** significant <u>digits</u>.)

(b) 4.0×10^{-9} mol/L
Because the pH of the solution is 8.40, $\log[H^+] = -8.\underline{40}$, we write the $[H^+] = 10^{-8.40}$.

Use your calculator to determine the antilogarithm of -8.40 as 3.98×10^{-9} (See Appendix II in your text).

$[H^+] = 3.98 \times 10^{-9}$ mol/L = **4.0** $\times 10^{-9}$ mol/L

(The antilogarithm of a number with <u>two</u> digits to the <u>right</u> of the decimal point will have **two** significant <u>digits</u>.)

If your answers are correct, proceed to Problem 15-11. If your answers are incorrect, review Section 15-5 (Examples 15-7 and 15-8 and Study Exercise 15-11) in your text.

Proceed to Problem 15-11.

PROBLEM 15-11

Complete and balance the following equations, writing them as total ionic equations and as net ionic equations; indicate any precipitate by (s) and any gas by (g). (You may use the periodic table, the electromotive or activity series, and the rules for the solubility of inorganic substances in water.)

(a) H_2SO_4 (aq) + Sr(NO_3)$_2$ (aq) --->

(b)   Ni(NO_3)$_2$ (aq) + H_2S (aq) --->

(c) Na_3PO_4 (aq) + Fe(NO_3)$_3$ (aq) --->

(d)   Zn (s) + HCl (aq) --->

ANSWERS AND SOLUTIONS TO PROBLEM 15-11

(a)   Completing and balancing the equation according to guideline 1 in your text gives

H_2SO_4 (aq) + Sr(NO_3)$_2$ (aq) ---> $SrSO_4$ (s) + 2 HNO_3 (aq)

Consider each of the reactants and products in regard to complete (molecular) or ionic form:

Formula	Identification	Conclusion
H_2SO_4	Strong acid	Ionic form
Sr(NO_3)$_2$	Soluble salt	Ionic form
$SrSO_4$	Insoluble salt (see solubility rules - back of book)	Complete form
HNO_3	Strong acid	Ionic form

We write the total ionic equation by applying guidelines 2, 3, and 4. The total ionic equation is

$$2\ H^+(\underline{aq}) + SO_4^{2-}(\underline{aq}) + Sr^{2+}(\underline{aq}) + 2\ NO_3^-(\underline{aq}) \longrightarrow$$

$$SrSO_4(\underline{s}) + 2\ H^+(\underline{aq}) + 2\ NO_3^-(\underline{aq})$$

Next, we check each ion and charge according to guideline 5.

$$2\ \overset{\checkmark}{H^+}(\underline{aq}) + \overset{\checkmark}{SO_4^{2-}}(\underline{aq}) + \overset{\checkmark}{Sr^{2+}}(\underline{aq}) + 2\ \overset{\checkmark}{NO_3^-}(\underline{aq}) \longrightarrow$$

charges: $2(1^+) \quad + 2^- \qquad\qquad + 2^+ \qquad\qquad + 2(1^-) \quad = \quad 0 =$

$$\overset{\checkmark\checkmark}{SrSO_4}(\underline{s}) + 2\ \overset{\checkmark}{H^+}(\underline{aq}) + 2\ \overset{\checkmark}{NO_3^-}(\underline{aq})$$

$$0 \qquad\qquad + 2(1^+) \qquad\qquad + 2(1^-) \qquad\quad = \quad 0$$

We can now write the net ionic equation by crossing out ions that appear on both sides of the equation, according to guideline 6. Finally, we check the net ionic equation for ions, charge, and lowest possible ratio of coefficients.

$$\cancel{2\ H^+}(\underline{aq}) + SO_4^{2-}(\underline{aq}) + Sr^{2+}(\underline{aq}) + \cancel{2\ NO_3^-}(\underline{aq}) \longrightarrow$$

$$SrSO_4(\underline{s}) + \cancel{2\ H^+}(\underline{aq}) + \cancel{2\ NO_3^-}(\underline{aq})$$

The net ionic equation is

$$\overset{\checkmark}{Sr^{2+}}(\underline{aq}) + \overset{\checkmark}{SO_4^{2-}}(\underline{aq}) \longrightarrow \overset{\checkmark\checkmark}{SrSO_4}(\underline{s})$$

charges: $\quad 2^+ \qquad\quad + 2^- \qquad = 0 = \quad 0$

(b) $Ni(NO_3)_2(\underline{aq}) + H_2S(\underline{aq}) \longrightarrow NiS(\underline{s}) + 2\ HNO_3(\underline{aq})$

Formula	Identification	Conclusion
$Ni(NO_3)_2$	Soluble salt	Ionic form
H_2S	Weak acid	Complete form
NiS	Insoluble salt (see solubility rules)	Complete form
HNO_3	Strong acid	Ionic form

The total ionic equation is

$$\overset{\checkmark}{Ni^{2+}}(\underline{aq}) + 2\ \overset{\checkmark}{NO_3^-}(\underline{aq}) + \overset{\checkmark\checkmark}{H_2S}(\underline{aq}) \longrightarrow$$

$$\overset{\checkmark\checkmark}{NiS}(\underline{s}) + 2\ \overset{\checkmark}{H^+}(\underline{aq}) + 2\ \overset{\checkmark}{NO_3^-}(\underline{aq})$$

charges: $2^+ + 2(1^-) \quad + 0 \quad = 0 = 0 \qquad + 2(1^+) \qquad + 2(1^-) = 0$

248

By crossing out ions that appear on both sides of the equation, we can write the net ionic equation:

$Ni^{2+}(aq) + \cancel{2NO_3^-}(aq) + H_2S(aq) \longrightarrow$

$$NiS(s) + 2H^+(aq) + \cancel{2NO_3^-}(aq)$$

$$Ni^{2+}(aq) + H_2S(aq) \longrightarrow NiS(s) + 2\ H^+(aq)$$

charges: 2^+ $+\ 0$ $=\ 2^+ = 0$ $+\ 2(1^+)$ $=\ 2^+$

(c) $Na_3PO_4(aq) + Fe(NO_3)_3(aq) \longrightarrow 3\ NaNO_3(aq) + FePO_4(s)$

Formula	Identification	Conclusion
Na_3PO_4	Soluble salt	Ionic form
$Fe(NO_3)_3$	Soluble salt	Ionic form
$NaNO_3$	Soluble salt (see solubility rules)	Ionic form
$FePO_4$	Insoluble salt (see solubility rules)	Complete form

The total ionic equation is

$$3\ Na^+(aq) + PO_4^{3-}(aq) + Fe^{3+}(aq) + 3\ NO_3^-(aq) \longrightarrow$$

charges: $3(1^+) + 3^-$ $+\ 3^+$ $+\ 3(1^-)$ $=\ 0$

$$3\ Na^+(aq) + 3\ NO_3^-(aq) + FePO_4(s)$$

$$=\ 3(1^+)\qquad +\ 3(1^-)\qquad +\ 0\qquad =\ 0$$

By crossing out ions that appear on both sides of the equation, we can write the net ionic equation:

$\cancel{3\ Na^+}(aq) + PO_4^{3-}(aq) + Fe^{3+}(aq) + \cancel{3\ NO_3^-}(aq) \longrightarrow$

$$\cancel{3\ Na^+}(aq) + \cancel{3\ NO_3^-}(aq) + FePO_4(s)$$

$$PO_4^{3-}(aq) + Fe^{3+}(aq) \longrightarrow FePO_4(s)$$

charges: 3^- $+\ 3^+$ $=\ 0\ =\ 0$

(d) $Zn(s) + 2\ HCl(aq) \longrightarrow ZnCl_2(aq) + H_2(g)$

This is a single-replacement reaction, involving the electromotive or activity series. Zinc is higher in the series than hydrogen so zinc will replace hydrogen from its acid.

Formula	Identification	Conclusion
Zn	Solid metal	Complete form
HCl	Strong acid	Ionic form
$ZnCl_2$	Soluble salt (see solubility rules)	Ionic form
H_2	Gas	Complete form

The total ionic equation is

$$Zn(\underline{s}) + 2\ H^+(\underline{aq}) + 2\ Cl^-(\underline{aq}) \longrightarrow Zn^{2+}(\underline{aq}) + 2\ Cl^-(\underline{aq}) + H_2(\underline{g})$$

charges: $0 + 2(1^+) + 2(1^-) = 0 = 2^+ + 2(1^-) + 0 = 0$

By crossing out ions that appear on both sides of the equation, we can write the net ionic equation:

$$Zn(\underline{s}) + 2\ H^+(\underline{aq}) + \cancel{2\ Cl^-}(\underline{aq}) \longrightarrow Zn^{2+}(\underline{aq}) + \cancel{2\ Cl^-}(\underline{aq}) + H_2(\underline{g})$$

(Note that neither Zn nor H^+ can be crossed out since they appear in the products as Zn^{2+} and H_2, respectively.)

$$Zn(\underline{s}) + 2\ H^+(\underline{aq}) \longrightarrow Zn^{2+}(\underline{aq}) + H_2(\underline{g})$$

charges: $0 + 2(1^+) = 2^+ = 2^+ + 0 = 2^+$

If your answers are correct, you have completed this chapter. If your answers are incorrect, review Sections 15-7 and 15-8 (Examples 15-9, 15-10, 15-11, 15-12, and 15-13 and Study Exercise 15-12) in your text.

These problems conclude the chapter on acids, bases, and ionic equations.

Now take the sample quizzes to see if you have mastered the material in this chapter.

SAMPLE QUIZZES

The quizzes divide this chapter into two parts. Quiz 15A (Sections 15-1 to 15-4) and Quiz 15B (Sections 15-5 to 15-8).

Quiz #15A (Sections 15-1 to 15-4)

You may use the periodic table.

1. Classify the folowing as an acid, base, or amphoteric sub-
    stance according to the Brønsted-Lowry theory.

    (a) HCO_3^-(aq) _____ ; (b) Cl^-(aq) _____

    (c) $H_2S_2O_3$(aq) _____ ; (d) $H_2PO_4^-$(aq) _____

2. If 24.8 mL of 0.285 M sodium hydroxide (NaOH) is necessary
    to neutralize 30.0 mL of a hydrochloric and (HCl) solution
    to a phenolphthalein end point, calculate the molarity of
    the hydrochloric acid solution.

3. If 0.786 g of sodium carbonate (Na_2CO_3) is dissolved in
    water and the solution titrated with 37.4 mL of hydrochlo-
    ric acid (HCl) to a methyl orange end point, calculate the
    molarity of the hydrochloric acid solution.   (Na = 23.0
    amu, C = 12.0 amu, O = 16.0 amu)

4. If 1.75 g of sodium hydroxide (NaOH) is titrated with 12.7
    mL of hydrochloric acid (HCl) to a phenolphthalein end
    point.   Calculate (a) the molarity and (b) the percent by
    mass (density of solution = 1.058 g/mL of the hydrochloric
    acid solution)(Na = 23.0 amu, O = 16.0 amu, H = 1.0 amu, Cl
    = 35.5 amu)

5. In the titration of 32.6 mL of sodium hydroxide solution of
    unknown concentration, 23.6 mL of 0.110 N sulfuric acid
    solution was required to neutralize the sodium hydroxide in
    reactions where both hydrogen ions of sulfuric acid react.
    Calculate the normality of the sodium hydroxide solution.

Solutions and Answers for Quiz #15A

1. (a) amphoteric; (b) base; (c) acid; (d) amphoteric

2. $NaOH(aq) + HCl(aq) \longrightarrow NaCl(aq) + H_2O(\ell)$

$$24.8 \text{ mL solution} \times \frac{1 \text{ L solution}}{1000 \text{ mL solution}} \times \frac{0.285 \text{ mol NaOH}}{1 \text{ L solution}}$$

$$\times \frac{1 \text{ mol HCl}}{1 \text{ mol NaOH}} = 0.00707 \text{ mol HCl}$$

$$\frac{0.00707 \text{ mol HCl}}{30.0 \text{ mL solution}} \times \frac{1000 \text{ mL solution}}{1 \text{ L solution}} = \frac{0.236 \text{ mol HCl}}{1 \text{ L solution}}$$

$$= 0.236 \text{ M}$$

3. $Na_2CO_3(s) + 2 HCl(aq) \longrightarrow 2 NaCl(aq) + H_2O(\ell) + CO_2(g)$

$$0.786 \text{ g Na}_2\text{CO}_3 \times \frac{1 \text{ mol Na}_2\text{CO}_3}{106.0 \text{ g Na}_2\text{CO}_3} \times \frac{2 \text{ mol HCl}}{1 \text{ mol Na}_2\text{CO}_3} = 0.0148 \text{ mol HCl}$$

$$\frac{0.0148 \text{ mol HCl}}{37.4 \text{ mL solution}} \times \frac{1000 \text{ mL}}{1 \text{ L}} = 0.396 \text{ M}$$

4. $NaOH(s) + HCl(aq) \longrightarrow NaCl(aq) + H_2O(\ell)$

(a) $$1.75 \text{ g NaOH} \times \frac{1 \text{ mol NaOH}}{40.0 \text{ g NaOH}} \times \frac{1 \text{ mol HCl}}{1 \text{ mol NaOH}} = 0.0438 \text{ mol HCl}$$

$$\frac{0.0438 \text{ mol HCl}}{12.7 \text{ mL solution}} \times \frac{1000 \text{ mL solution}}{1 \text{ L solution}} = \frac{3.45 \text{ mol HCl}}{1 \text{ L solution}}$$

$$= 3.45 \text{ M}$$

(b) $$\frac{3.45 \text{ mol HCl}}{1 \text{ L solution}} \times \frac{36.5 \text{ g HCl}}{1 \text{ mol HCl}} \times \frac{1 \text{ L}}{1000 \text{ mL}} \times \frac{1 \text{ mL solution}}{1.058 \text{ g solution}}$$

$$\times 100 = 11.9 \text{ \%}$$

5. $V_A \cdot N_A = V_B \cdot N_B$

$$0.0236 \text{ L H}_2\text{SO}_4 \text{ solution} \times \frac{0.110 \text{ eq H}_2\text{SO}_4}{\text{L H}_2\text{SO}_4 \text{ solution}}$$

$$= 0.0326 \text{ L NaOH solution} \times N_B$$

$$N_B = \frac{(0.0236 \text{ L})(0.110 \text{ eq/L})}{(0.0326 \text{ L})} = \frac{0.0796 \text{ eq NaOH}}{\text{L NaOH solution}} = 0.0796 \text{ N}$$

Quiz #15B (Sections 15-5 to 15-8)

You may use the periodic table, the rules for the solubility of inorganic substances in the water, and the electromotive or activity series.

1. Calculate the pH and pOH of the following solutions:

(a) $[H^+] = 3.0 \times 10^{-6}$ mol/L

(b) $[H^+] = 6.4 \times 10^{-3}$ mol/L

2. Calculate the hydrogen ion concentration in moles per liter for each of the following solutions:

(a) pH = 3.60

(b) pH = 5.50

3. Complete and balance the following equations, writing them as total ionic equations and as net ionic equations. Indicate any precipitate by (s) and any gas by a (g).

(a) $CaCl_2(\underline{aq}) + (NH_4)_2CO_3(\underline{aq})$ --->

(b) $Pb^{2+}(\underline{aq}) + H_2S(\underline{aq})$ --->

(c) $Mg(\underline{s}) + HCl(\underline{aq})$ --->

Solutions and Answers for Quiz #15B

1. (a) pH = -[-5.5229], 5.52

   pOH = 14.00 - 5.52 = 8.48

   (b) pH = -[-2.1938], 2.19

   pOH = 14.00 - 2.19 = 11.81

2. (a) $[H^+] = 10^{-3.60} = 2.5 \times 10^{-4}$ mol/L

   (b) $[H^+] = 10^{-5.50} = 3.2 \times 10^{-6}$ mol/L

3. (a) Ionic: $Ca^{2+}(\underline{aq}) + 2\ Cl^-(\underline{aq}) + 2\ NH_4^+(\underline{aq}) + CO_3^{2-}(\underline{aq})$ --->

   $CaCO_3(\underline{s}) + 2\ NH_4^+(\underline{aq}) + 2\ Cl^-(\underline{aq})$

   Net: $Ca^{2+}(\underline{aq}) + CO_3^{2-}(\underline{aq})$ ---> $CaCO_3(\underline{s})$

   (b) Ionic and net: $Pb^{2+}(\underline{aq}) + H_2S(\underline{aq})$ ---> $PbS(\underline{s}) + 2\ H^+(\underline{aq})$

   (c) Ionic: $Mg(\underline{s}) + 2\ H^+(\underline{aq}) + 2\ Cl^-(\underline{aq})$ --->

   $Mg^{2+}(\underline{aq}) + 2\ Cl^-(\underline{aq}) + H_2(\underline{g})$

   Net: $Mg(\underline{s}) + 2\ H^+(\underline{aq})$ ---> $Mg^{2+}(\underline{aq}) + H_2(\underline{g})$

Review Exam #5 (Chapters 13 to 15) [Answers are in ( ) to the right of each question.]

You may use the periodic table, the rules for the solubility of inorganic substances in water, and the electromotive or activity series.

Element	Atomic Mass Units (amu)
Na	23.0
I	126.9
H	1.0
O	16.0
Li	6.9
S	32.1
C	12.0

1. Complete and balance the following equation; indicate any precipitate by (s) and any gas by (g).

$$Al(\underline{s}) + H_2O(\underline{g}) \xrightarrow{\triangle}$$

A. $2\ Al(\underline{s}) + 3\ H_2O(\underline{g}) \xrightarrow{\triangle} Al_2O_3(\underline{s}) + 6\ H(\underline{g})$ (D)

B. $2\ Al(\underline{s}) + 2\ H_2O(\underline{g}) \xrightarrow{\triangle} 2\ Al(OH)(\underline{s}) + H_2(\underline{g})$

C. $2\ Al(\underline{s}) + 6\ H_2O(\underline{g}) \xrightarrow{\triangle} 2\ Al(OH)_3(\underline{s}) + 3\ H_2(\underline{g})$

D. $2\ Al(\underline{s}) + 3\ H_2O(\underline{g}) \xrightarrow{\triangle} Al_2O_3(\underline{s}) + 3\ H_2(\underline{g})$

E. $Al(\underline{s}) + H_2O(\underline{g}) \xrightarrow{\triangle} AlO(\underline{s}) + H_2(\underline{g})$

2. Complete and balance the following equation:

$$BaO(\underline{s}) + HCl(\underline{aq}) \dashrightarrow$$

A. $BaO(\underline{s}) + HCl(\underline{aq}) \dashrightarrow BaCl(\underline{aq}) + HO(\ell)$ (B)

B. $BaO(\underline{s}) + 2\ HCl(\underline{aq}) \dashrightarrow BaCl_2(\underline{aq}) + H_2O(\ell)$

C. $BaO(\underline{s}) + HCl(\underline{aq}) \dashrightarrow BaClO(\underline{aq}) + H(\ell)$

D. $2\ BaO(\underline{s}) + 2\ HCl(\underline{aq}) \dashrightarrow 2\ BaClO(\underline{aq}) + H_2(\underline{g})$

E. $2\ BaO(\underline{s}) + 2\ HCl(\underline{sq}) \dashrightarrow 2\ BaCl(\underline{aq}) + H_2O_2(\ell)$

3. Calculate the percent water in sodium iodide dihydrate.

A. 19.4 %   (A)

B. 9.68 %

C. 10.7 %

D. 21.4 %

E. 28.4 %

4. Calculate the formula of the hydrate of lithium iodide containing 28.8 % water.

A. $LiI_2 \cdot 2\ H_2O$ (E)

B. $LiI_2 \cdot H_2O$

C. $LiI \cdot H_2O$

D. $LiI \cdot 2\ H_2O$

E. $LiI \cdot 3\ H_2O$

5. Calculate the percent by mass sodium iodide if 18.6 g of sodium iodide is dissolved in 120.0 g of water.

   A.   18.6 %   (E)

   B.   15.5 %

   C. $12\overline{0}$ %

   D.   12.0 %

   E.   13.4 %

6. Calculate the number of grams of sodium iodide that must be dissolved in 225 g of water in the preparation of a 12.5 % by mass sodium iodide solution.

   A.   12.5 g   (B)

   B.   32.1 g

   C.   28.1 g

   D.   22.5 g

   E.   15.8 g

7. Calculate the number of milligrams of potassium ions (K^+) in 2.60 L of a water sample having 200 parts per million (ppm). (Assume that the density of the very dilute water sample is 1.00 g/mL.)

   A.   0.520 mg   (D)

   B.   5.20 mg

   C.   52.0 mg

   D. $52\overline{0}$ mg

   E.   13.0 mg

8. Calculate the molarity of sodium iodide solution that contains 15.6 g sodium iodide in $4\overline{0}0$ mL of solution.

   A.   0.210 $\underline{M}$ (C)

   B.   24.0 $\underline{M}$

   C.   0.260 $\underline{M}$

   D.   0.104 $\underline{M}$

   E.   39.0 $\underline{M}$

9. Calculate the number of grams of sodium iodide necessary to prepare 280 mL of a 0.600 $\underline{M}$ sodium iodide solution.

    A. 3.86 g   (D)

    B. 112 g

    C. 168 g

    D. 25.2 g

    E. 21.3 g

10. Calculate the normality of a solution containing 67.0 g of sulfuric acid in 250 mL of solution in reactions that replace both hydrogen ions.

    A. 5.47 $\underline{N}$ (A)

    B. 2.73 $\underline{N}$

    C. 1.37 $\underline{N}$

    D. 0.683 $\underline{N}$

    E. 2.00 $\underline{N}$

11. Calculate the number of milliliters of a 2.00 $\underline{N}$ sulfuric acid solution necessary to provide 85.0 g of sulfuric acid in reactions that replace both hydrogen ions.

    A. 433 mL   (B)

    B. 867 mL

    C. 1,710 mL

    D. 2,000 mL

    E. 1,150 mL

12. Calculate the molality of a sulfuric acid solution containing 85.0 g sulfuric acid in 650 g of water.

    A. 0.866 $\underline{m}$ (E)

    B. 13.1 $\underline{m}$

    C. 2.67 $\underline{m}$

    D. 1.18 $\underline{m}$

    E. 1.33 $\underline{m}$

13. Calculate the number of grams of water that must be added to 75.0 g of sulfuric acid to prepare a 1.50 $\underline{m}$ sulfuric acid solution.

   A. 1,5$\overline{0}$0 g   (E)

   B. 1,020 g

   C. 1,310 g

   D. 1,150 g

   E. 51$\overline{0}$ g

14. Calculate the molarity of a 0.860 $\underline{N}$ sulfuric acid solution in reactions that replace both hydrogen ions.

   A. 1.72 $\underline{M}$ (B)

   B. 0.430 $\underline{M}$

   C. 84.4 $\underline{M}$

   D. 42.1 $\underline{M}$

   E. 0.860 $\underline{M}$

15. Calculate the normality of a 6.00 $\underline{M}$ sulfuric acid solution in reactions that replace both hydrogen ions.

   A. 1.00 $\underline{N}$ (E)

   B. 2.00 $\underline{N}$

   C. 3.00 $\underline{N}$

   D. 6.00 $\underline{N}$

   E. 12.0 $\underline{N}$

16. Calculate the freezing point of an aqueous solution containing 12.5 g of sugar ($C_{12}H_{22}O_{11}$) in 85.0 g of water.

   K_f = 1.86°C/$\underline{m}$, freezing point of water = 0.00°C)

   A. -0.80°C   (A)

   B. -0.93°C

   C. -0.43°C

   D. -0.86°C

   E. 0.86°C

17. The substance ethylamine solution ($C_2H_5-NH_2$) would be classified as:

   A. salt   (E)

   B. strong acid

   C. strong base

   D. weak acid

   E. weak base

18. The polyatomic ion, HCO_3^-, would be classified as a Brønsted-Lowry

   A. acid   (C)

   B. base

   C. amphoteric substance

19. If 35.30 mL of a 0.300 $\underline{M}$ sodium hydroxide solution is necessary to neutralize 5.00 mL of a solution of hydrochloric acid to a phenolphthalein end point, calculate the molarity of the hydrochloric acid solution.

   A. 3.00 $\underline{M}$ (D)

   B. 0.300 $\underline{M}$

   C. 0.0426 $\underline{M}$

   D. 2.12 $\underline{M}$

   E. 0.212 $\underline{M}$

20. If 0.625 g of pure sodium carbonate is titrated with 10.40 mL of hydrochloric acid solution to a methyl orange end point, calculate the molarity of the hydrochloric acid solution. (Hint: The methyl orange end point takes the carbonate ion to carbon dioxide.)

   A. 1.13 $\underline{M}$ (A)

   B. 0.567 $\underline{M}$

   C. 1.45 $\underline{M}$

   D. 0.724 $\underline{M}$

   E. 0.120 $\underline{M}$

21. In the titration of 29.50 mL of sodium hydroxide solution of unknown concentration, 36.20 mL of 0.130 $\underline{N}$ sulfuric acid solution was required to neutralize the sodium hydroxide in reactions where both hydrogen ions of sulfuric acid react. Calculate the normality of the sodium hyroxide solution.

    A. 0.0530 $\underline{N}$ (C)

    B. 0.106 $\underline{N}$

    C. 0.160 $\underline{N}$

    D. 0.320 $\underline{N}$

    E. 0.0800 $\underline{N}$

22. Calculate the pH of a solution whose hydrogen ion concentration is 1.6×10^{-9} mol/L.

    A. 9.00   (D)

    B. 6.20

    C. 7.80

    D. 8.80

    E. 5.20

23. Calculate the hydrogen ion concentration in moles per liter for a solution whose pH is 5.60.

    A. 11.5×10^{-6} mol/L   (B)

    B. 2.5×10^{-6} mol/L

    C. 4.0×10^{-6} mol/L

    D. 7.5×10^{-5} mol/L

    E. 7.5×10^{-7} mol/L

24. Complete and balance the following equation, writing it as a total ionic equation and as a net ionic equation; indicate any precipitate by (s) and any gas by (g):

$$Pb(NO_3)_2(aq) + HCl(aq) \xrightarrow{\text{cold}}$$

The balanced <u>net</u> ionic equation is as follows:

A. $Pb(NO_3)_2(aq) + 2\ Cl^-(aq) \xrightarrow{\text{cold}}$ (E)

$$PbCl_2(s) + 2\ NO_3^-(aq)$$

B. $Pb^{2+}(aq) + 2\ NO_3^-(aq) + 2\ Cl^-(aq) \xrightarrow{\text{cold}}$

$$PbCl_2(s) + 2\ NO_3^-(aq)$$

C. $Pb^{2+}(aq) + 2\ H^+(aq) + 2\ Cl^-(aq) \xrightarrow{\text{cold}}$

$$PbCl_2(s) + 2\ H^+(aq)$$

D. $Pb^{2+}(aq) + 2\ NO_3^-(aq) + 2\ H^+(aq) \xrightarrow{\text{cold}}$

$$PbCl_2(s) + 2\ H^+(aq) + 2\ NO_3^-(aq)$$

E. $Pb^{2+}(aq) + 2\ Cl^-(aq) \xrightarrow{\text{cold}} PbCl_2(s)$

25. Complete and balance the following equation, writing it as a total ionic equation and as a net ionic equation; indicate any precipitate by (s) and any gas by (g):

$$Cd(s) + HCl(aq) \longrightarrow$$

The balanced <u>net</u> equation is as follows:

A. $Cd(s) + 2\ H^+(aq) + 2\ Cl^-(aq) \longrightarrow$

$$Cd^{2+}(aq) + 2\ Cl^-(aq) + H_2(g)$$ (B)

B. $Cd(s) + 2\ H^+(aq) \longrightarrow Cd^{2+}(aq) + H_2(g)$

C. $Cd(s) + 2\ HCl(aq) \longrightarrow Cd^{2+}(aq) + 2\ Cl^-(aq) + H_2(g)$

D. $Cd(s) + 2\ HCl(aq) \longrightarrow Cd^{2+}(aq) + 2\ Cl^-(aq) + 2\ H^+(aq)$

E. $Cd(s) + 2\ Cl^-(aq) \longrightarrow CdCl_2(s)$

CHAPTER 16

OXIDATION-REDUCTION EQUATIONS

In Chapter 9 we considered the five simple types of reactions (combination, decomposition, single replacement, double replacement, and neutralization), and discussed balancing the equations "by inspection." In this chapter we will consider another type of reaction -- an oxidation-reduction reaction. An <u>oxidation-reduction</u> (<u>redox</u>) <u>reaction</u> is a type of chemical reaction in which one substance transfers electrons to another substance. To balance these complex oxidation-reduction equations we need to use special techniques which will be considered in this chapter.

SELECTED TOPICS

1. <u>Oxidation</u> is any chemical reaction in which a substance loses electrons. <u>Reduction</u> is any chemical reaction in which a substance gains electrons. The substance being oxidized is called the <u>reducing agent</u> (<u>reductant</u>) because it produces reduction in another substance. The substance being reduced is called the <u>oxidizing agent</u> (<u>oxidant</u>) because it produces oxidation in another substance. <u>Oxidation always accompanies reduction</u> and <u>reduction always accompanies oxidation</u>. One way of remembering these definitions is that <u>loss</u> of <u>electrons</u> is OXIDATION (<u>l</u> <u>O</u> <u>e</u>) and <u>gain</u> of <u>electrons</u> is REDUCTION (<u>g</u> <u>R</u> <u>e</u>), see Figure 16-1 in your text. See Section 16-1.

2. In Chapter 6 (Section 6-3, in your text) we defined an <u>oxidation</u> <u>number</u> (<u>ox. no. or oxidation state</u>) as a positive or negative whole number assigned to an element in a compound or ion. It is based on certain rules. These rules are given in Section 6-3 in your text and reviewed below:

<u>Rule</u>	<u>State of Element</u>	<u>Oxidation Number</u> (ox no)	<u>Example</u>
<u>1</u>	Neutral compound	Sum of ox nos = 0	$NaCl$: $(+1) + (-1) = 0$
<u>2</u>	Uncombined	Zero	Na^0: ox no = 0
<u>3</u>	Monatomic ion	Same as ionic charge	Na^+: ox no = +1
<u>4</u>	Polyatomic ion	Its ionic charge = sum of ox nos on all atoms	ClO^-: +1 (Cl) + [-2(O)] = -1
<u>5</u>	Metals combined with nonmetals Nonmetals combined with metals	Positive Negative	$NaCl$: Na^+ +1 Cl^- -1

262

Rule	State of Element	Oxidation Number	Example
6	Two nonmetals combined	More electronegative atom - negative	NO: N +2; O -2
7	Hydrogen	Usually +1 Except hydrides -1	HCl: H +1 NaH: H -1
8	Oxygen	Usually -2 Except peroxides -1	MgO: O -2 H_2O_2: O -1

You will need to calculate oxidation numbers in one of the special techniques for balancing oxidation-reduction equations. See Section 16-2 in your text.

3. The two methods for balancing oxidation-reduction equations are: (1) the oxidation number method, and (2) the ion electron method. We will consider both. In the oxidation number method we calculate the oxidation numbers of the elements in the compounds that undergo changes in oxidation number. Section 16-3 gives guidelines in balancing oxidation-reduction equations by the oxidation number method. Study these guidelines. See Section 16-3 in your text.

4. The ion electron method uses two partial equations representing half-reactions. One half-reaction describes the oxidation and the other equation describes the reduction. Although we artifically divide the equations into two partial equations, these partial equations do not take place alone. Whenever oxidation occurs, so does reduction. Section 16-4 gives guidelines for balancing oxidation-reduction equations by the ion electron method. Study these

   guidelines. Note in guideline 3 we must add H^+ or H_2O as needed to balance H or O atoms on each side of the half-reaction in acid solution. In basic solution we must add

   H_2O or OH^- to balance H or O atoms on each side of the half-reaction.

In acid:	Need	Add	
	H	H^+	
	O	H_2O	---> 2 H^+
In base:	Need	Add	
	H	H_2O	---> OH^-
	O	2 OH^-	---> H_2O

See Section 16-4 in your text.

PROBLEM 16-1

Calculate the oxidation number for the element indicated in each of the following compounds or ions:

(a)   As in As_2O_3 (a) _____

(b)   As in $HAsO_3$ (b) _____

(c)   As in As_2O_5 (c) _____

(d)   As in $AsO_4{}^{3-}$ (d) _____

(e)   As in $AsO_2{}^{-}$ (e) _____

ANSWERS AND SOLUTIONS TO PROBLEM 16-1

(a)   +3
$$2(\text{ox. no. As}) + 3(-2) = 0$$
$$2(\text{ox. no. As}) - 6 = 0$$
$$\text{ox. no. As} = +3$$

(b)   +5
$$+1 + \text{ox. no. As} + 3(-2) = 0$$
$$+1 + \text{ox. no. As} - 6 = 0$$
$$\text{ox. no. As} - 5 = 0$$
$$\text{ox. no. As} = +5$$

(c)   +5
$$2(\text{ox. no. As}) + 5(-2) = 0$$
$$2(\text{ox. no. As}) - 10 = 0$$
$$\text{ox. no. As} = +5$$

(d)   +5
$$\text{ox. no. As} + 4(-2) = -3$$
$$\text{ox. no. As} - 8 = -3$$
$$\text{ox. no. As} = 8 - 3$$
$$\text{ox. no. As} = +5$$

(e)   +3
$$\text{ox. no. As} + 2(-2) = -1$$
$$\text{ox. no. As} - 4 = -1$$
$$\text{ox. no. As} = 4 - 1$$
$$\text{ox. no. As} = +3$$

If your answers are correct, proceed to Problem 16-2. If your answers are incorrect, review Section 6-3 (Example 6-1 and Study Exercise 6-1) and Section 16-2 (Example 16-1 and Study Exercise 16-1) in your text.

Proceed to Problem 16-2

PROBLEM 16-2

(a) Balance the following oxidation-reduction equations by the oxidation number method.

(b) Determine the substances that are oxidized and reduced and the oxidizing and reducing agents.

   (1) H_3SbO_4 + HI ---> H_3SbO_3 + H_2O + I_2

   (2) $AsCl_3$ + I_2 + HCl ---> $AsCl_5$ + HI

   (3) $Ba(BrO_3)_2$ + H_3AsO_3 ---> $BaBr_2$ + H_3AsO_4

ANSWERS AND SOLUTIONS TO PROBLEM 16-2

(1) (a) H_3SbO_4 + HI ---> H_3SbO_3 + H_2O + I_2

      According to guideline 1, determine by inspection the elements undergoing a change in oxidation number. These elements are Sb and I. Following guideline 2, write the oxidation number above the element oxidized or reduced.

$$\overset{5^+}{H_3SbO_4} + \overset{1^-}{HI} \longrightarrow \overset{3^+}{H_3SbO_3} + H_2O + \overset{0}{I_2}$$

Determine the number of electrons lost or gained for each element that undergoes a change in oxidation number, according to guideline 3. Then balance the number of electrons lost and the number of electrons gained by placing a coefficients in front of the loss and gain of electrons, according to guideline 4.

$$\overset{1(\text{gain } 2\text{ e}^-)}{\overset{5^+}{H_3SbO_4} + \overset{1^-}{HI} \longrightarrow \overset{3^+}{H_3SbO_3} + H_2O + \overset{0}{I_2}}$$
$$2(\text{loss } 1\text{ e}^-)$$

Place these coefficients in front of the corresponding formulas and balance the equation by inspection, according to guideline 5.

$$\overset{(\text{gain } 2\text{ e}^-)}{\overset{5^+}{H_3SbO_4} + 2\overset{1^-}{HI} \longrightarrow \overset{3^+}{H_3SbO_3} + H_2O + \overset{0}{I_2}}$$
$$2(\text{loss } 1\text{ e}^-)$$

(Because 1 is understood in front of H_3SbO_4, we can delete it.) Check each atom on both sides of the equation according to guideline 6.

$$\checkmark\checkmark\checkmark \quad \checkmark\checkmark \quad \checkmark\checkmark\checkmark \quad \checkmark\checkmark \quad \checkmark$$
$$H_3SbO_4 + 2 HI \longrightarrow H_3SbO_3 + H_2O + I_2$$

(b) HI - oxidized (I in HI loses electrons)
HI - reducing agent

H_3SbO_4 - reduced (Sb in H_3SbO_4 gains electrons)
H_3SbO_4 - oxidizing agent

(2) (a) $AsCl_3 + I_2 + HCl \longrightarrow AsCl_5 + HI$

Apply guidelines 1 and 2 to this equation and place the oxidation numbers above the elements that change in oxidation number.

$$\overset{3^+}{AsCl_3} + \overset{0}{I_2} + HCl \longrightarrow \overset{5^+}{AsCl_5} + \overset{1^-}{HI}$$

Following guideline 3, determine the number of electrons lost or gained for each element undergoing a change in oxidation number. Because there are 2 I atoms changing, the total gain in electrons is 2.

266

Therefore, we must balance with coefficients of 1 in both cases, according to guideline 4.

$$\overset{\overset{\displaystyle 1(\text{loss } 2\ e^-)}{\underline{\hspace{5cm}}}}{\underset{\underset{1(2 \times \text{gain } 1\ e^-)}{\overline{\hspace{5cm}}}}{\overset{3+}{As}Cl_3 + \overset{0}{I_2} + HCl \longrightarrow \overset{5+}{As}Cl_5 + H\overset{1-}{I}}}$$

Place the coefficients in front of the formulas and balance by inspection, according to guideline 5.

$$\overset{\overset{\displaystyle 1(\text{loss } 2\ e^-)}{\underline{\hspace{5cm}}}}{\underset{\underset{1(2 \times \text{gain } 1\ e^-)}{\overline{\hspace{5cm}}}}{\overset{3+}{As}Cl_3 + \overset{0}{I_2} + 2\ HCl \longrightarrow \overset{5+}{As}Cl_5 + 2\ H\overset{1-}{I}}}$$

Check each atom on both sides of the equation according to guideline 6.

$$\overset{\checkmark\checkmark}{As}Cl_3 + \overset{\checkmark}{I_2} + 2\ \overset{\checkmark\checkmark}{H}Cl \longrightarrow \overset{\checkmark\checkmark}{As}Cl_5 + 2\ \overset{\checkmark\checkmark}{H}I$$

(b) $AsCl_3$ - oxidized (As in $AsCl_3$ loses electrons)
$AsCl_3$ - reducing agent

I_2 - reduced (I decreases gains electrons)
I_2 - oxidizing agent

(3) (a) $Ba(BrO_3)_2 + H_3AsO_3 \longrightarrow BaBr_2 + H_3AsO_4$

Apply guidelines 1 and 2 to this equation and place the oxidation numbers above the elements that change in oxidation number.

$$Ba(\overset{5+}{Br}O_3)_2 + H_3\overset{3+}{As}O_3 \longrightarrow Ba\overset{1-}{Br}_2 + H_3\overset{5+}{As}O_4$$

Apply guideline 3 for a gain of 6 electrons for the Br and a loss of 2 electrons for As. Because there are 2 Br atoms changing in $Ba(BrO_3)_2$, the total gain in electrons is 12. Therefore we must balance with a coeffficient of 6 in front of the H_3AsO_3 according to guideline 4.

$$\overset{\overset{\displaystyle 1(2 \times \text{gain } 6\ e^- \text{ per atom})}{\underline{\hspace{5cm}}}}{\underset{\underset{6(\text{loss } 2\ e^-)}{\overline{\hspace{5cm}}}}{Ba(\overset{5+}{Br}O_3)_2 + 6\ H_3\overset{3+}{As}O_3 \longrightarrow Ba\overset{1-}{Br}_2 + H_3\overset{5+}{As}O_4}}$$

Place the coefficients in front of the formulas and balance the equation by inspection, according to guideline 5.

267

$$1(2 \times \text{gain } 6 \ e^- \text{ per atom})$$

$$\overset{5+}{Ba}(BrO_3)_2 + 6 \ \overset{3+}{H_3}AsO_3 \longrightarrow Ba\overset{1-}{Br}_2 + 6 \ \overset{5+}{H_3}AsO_4$$

$$6(\text{loss } 2 \ e^-)$$

Check each atom on both sides of the equation, according to guideline 6.

$$Ba(BrO_3)_2 + 6 \ H_3AsO_3 \longrightarrow BaBr_2 + 6 \ H_3AsO_4$$

(b) $Ba(BrO_3)_2$ - reduced [Br in $Ba(BrO_3)_2$ gains electrons]
$Ba(BrO_3)_2$ - oxidizing agent

H_3AsO_3 - oxidized (As in H_3AsO_3 loses electrons)
H_3AsO_3 - reducing agent

If your answers are correct, proceed to Problem 16-3. If your answers are incorrect, review Section 16-3 (Examples 16-2, 16-3, and 16-4 and Study Exercise 16-2) in your text.

Proceed to Problem 16-3.

PROBLEM 16-3

(a) Balance the following oxidation-reduction equations by the ion electron method.

(b) Determine the substances that are oxidized and reduced and the oxidizing and reducing agents.

(1) $ClO_3^- + I^- \longrightarrow Cl^- + I_2$ in acid solution

(2) $KBrO_3 + KI \longrightarrow KBr + I_2$ in acid solution

(3) $ClO_3^- + N_2H_4 \longrightarrow NO_3^- + Cl^-$ in basic solution

268

(1) (a) $ClO_3^- + I^- \dashrightarrow Cl^- + I_2$ in acid solution

Guideline 1 does not apply, because the equation is already in ionic form, so use guideline 2. The following two partial equations for the half-reactions result:

(1) $ClO_3^- \dashrightarrow Cl^-$

(2) $I^- \dashrightarrow I_2$

Balance the atoms for each partial equation, according to guideline 3. In partial equation (1) add 3 H_2O to the products to balance the O atoms in ClO_3^- and then add 6 H^+ to the reactants, because the reaction is carried out in acid.

(1) $ClO_3^- + 6\ H^+ \dashrightarrow Cl^- + 3\ H_2O$

(2) $2\ I^- \dashrightarrow I_2$

Next, balance the two partial equations electrically by adding electrons (electrons are negatively charged) to the appropriate sides, according to guideline 4. In partial equation (1), electrons are gained; hence, this equation represents the <u>reduction</u> half-reaction. In partial equation (2), electrons are lost; hence, this equation represents the <u>oxidation</u> half-reaction.

(1) Reduction: $ClO_3^- + 6\ H^+ + 6\ e^- \dashrightarrow Cl^- + 3\ H_2O$

  charges: $1^- + 6(1^+) + 6(1^-) = 1^- = 1^-$

(2) Oxidation: $2\ I^- \dashrightarrow I_2 + 2\ e^-$

  charges: $2(1^-) = 2^- = 2^-$

Multiply the entire partial equation (2) by 3, so that the gain of electrons in partial equation (1) is equal to the loss, according to guideline 5.

(1) Reduction: $ClO_3^- + 6\ H^+ + 6\ e^- \dashrightarrow Cl^- + 3\ H_2O$

(2) Oxidation: $6\ I^- \dashrightarrow 3\ I_2 + 6\ e^-$

Add the two partial equations and eliminate those electrons on both sides of the equation, according to guideline 6.

$$ClO_3^- + 6 \ H^+ + \cancel{6 \ e^-} + 6 \ I^- \ \text{---}> \ Cl^- + 3 \ H_2O + 3 \ I_2 + \cancel{6 \ e^-}$$

Check each atom and the charges on both sides of the equation to obtain the final equation, according to guideline 7.

$$\overset{\checkmark\checkmark}{ClO_3^-} + 6 \ \overset{\checkmark}{H^+} + 6 \ \overset{\checkmark}{I^-} \ \text{---}> \ \overset{\checkmark}{Cl^-} + 3 \ \overset{\checkmark\checkmark}{H_2O} + 3 \ \overset{\checkmark}{I_2}$$

charges: $1^- + 6(1^+) + 6(1^-) = 1^- = 1^-$

(b) ClO_3^- - reduced (the reduction half-reaction)

ClO_3^- - oxidizing agent

I^- - oxidized (the oxidation half-reaction)

I^- - reducing agent

(2)   (a) $KBrO_3 + KI \ \text{---}> \ KBr + I_2$ in acid solution

Apply guideline 1 by writing the equation in net ionic form without attempting to balance it (see Section 15-7 in your text. Refer to the solubility rules of inorganic substances in water in the back of your text.

$$K^+ + BrO_3^- + K^+ + I^- \ \text{---}> \ K^+ + Br^- + I_2$$

Net Ionic: $BrO_3^- + I^- \ \text{---}> \ Br^- + I_2$

Write two partial equations according to guideline 2:

$$BrO_3^- \ \text{---}> \ Br^-$$

$$I^- \ \text{---}> \ I_2$$

Balance the atoms for each partial equation, according to guideline 3:

(1) $BrO_3^- + 6 \ H^+ \ \text{---}> \ Br^- + 3 \ H_2O$

(2) $2 \ I^- \ \text{---}> \ I_2$

Then balance these two equations electrically, according to guideline 4. In partial equation (1) electrons are gained; hence, this equation represents the reduction half-reaction. In partial equation (2) electrons are lost; hence, this equation represents the oxidation half-reaction.

(1) Reduction: $BrO_3^- + 6 \ H^+ + 6 \ e^- \ \text{---}> \ Br^- + 3 \ H_2O$

charges: $1^- + 6(1^+) + 6^- = 1^- = 1^-$

(2) Oxidation: $2 I^- ---> I_2 + 2 e^-$

charges: $2(1^-) = 2^- = 2^-$

In the two partial equations, the number of electrons lost must be equal to the number of electrons gained, according to guideline 5; therefore, multiply partial equation (2) by 3.

(1) Reduction: $BrO_3^- + 6 H^+ + 6 e^- ---> Br^- + 3 H_2O$

(2) Oxidation: $6 I^- ---> 3 I_2 + 6 e^-$

Add the two partial equations and eliminate the electrons on opposite sides of the equation, according to guideline 6.

$$BrO_3^- + 6 H^+ + 6 I^- + \cancel{6 e^-} ---> Br^- + 3 H_2O + 3 I_2 + \cancel{6 e^-}$$

$$BrO_3^- + 6 H^+ + 6 I^- ---> Br^- + 3 H_2O + 3 I_2$$

Check each atom and charges on both sides of the equation to obtain the final balanced equation, according to guideline 7.

$$BrO_3^- + 6 H^+ + 6 I^- ---> Br^- + 3 H_2O + 3 I_2$$

charges: $1^- + 6(1^+) + 6^- = 1^- = 1^-$

(b) BrO_3^- - reduced (the reduction half-reaction)

BrO_3^- - oxidizing agent

I^- - oxidized (the oxidation half-reaction)

I^- - reducing agent

(3) (a) $ClO_3^- + N_2H_4 ---> NO_3^- + Cl^-$ in basic solution

Excluding guideline 1, because the equation is already in ionic form, and proceeding to guideline 2, we can write the following two partial equations for the half-reactions:

(1) $ClO_3^- ---> Cl^-$

(2) $N_2H_4 ---> NO_3^-$

Note that this reaction is carried out in a basic solution. Balance the atoms for each partial equation according to guideline 3:

271

(1) $$ClO_3^- + 3\ H_2O \longrightarrow Cl^- + 6\ OH^-$$

(2) $$N_2H_4 + 4\ OH^- + 12\ OH^- \longrightarrow 2\ NO_3^- + 4\ H_2O + 6\ H_2O$$

$$N_2H_4 + 16\ OH^- \longrightarrow 2\ NO_3^- + 10\ H_2O$$

Balance these two equations electrically, according to guideline 4. In partial equation (1), electrons are gained; hence, this equation represents the <u>reduction</u> half-reaction. In partial equation (2), electrons are lost; hence, this equation represents the <u>oxidation</u> half-reaction.

(1) Reduction: $ClO_3^- + 3\ H_2O + 6\ e^- \longrightarrow Cl^- + 6\ OH^-$

charges: $\quad 1^- \qquad + 6^- = 7^- = 1^- + 6^- = 7^-$

(2) Oxidation: $N_2H_4 + 16\ OH^- \longrightarrow 2\ NO_3^- + 10\ H_2O + 14\ e^-$

charges: $\qquad 16^- \quad = \quad 2(1^-) \qquad + 14^- = 16^-$

According to guideline 5, the number of electrons gained must equal the number of electrons lost, so multiply partial equation (1) by 7 and partial equation (2) by 3.

(1) Reduction: $7\ ClO_3^- + 21\ H_2O + 42\ e^- \longrightarrow 7\ Cl^- + 42\ OH^-$

charges: $\qquad 7^- \qquad + 42^- = 49^- = 7^- + 42^- = 49^-$

(2) Oxidation: $3\ N_2H_4 + 48\ OH^- \longrightarrow 6\ NO_3^- + 30\ H_2O + 42\ e^-$

charges: $\qquad 48^- \quad = \quad 6(1^-) \qquad + 42\ e^- = 48^-$

Add the two partial equations and eliminate the electrons, OH^- ions, and water molecules on both sides of the equation, according to guideline 6.

$$7\ ClO_3^- + \cancel{21\ H_2O} + \cancel{42\ e^-} + 3\ N_2H_4 + \overset{6}{\cancel{48}}\ OH^- \longrightarrow$$

$$7\ Cl^- + \cancel{42\ OH^-} + 6\ NO_3^- + \overset{9}{\cancel{30}}\ H_2O + \cancel{42\ e^-}$$

Check each atom and charge on both sides of the equation to obtain the final balanced equation, according to guideline 7.

$$7\ ClO_3^- + 3\ N_2H_4 + 6\ OH^- \longrightarrow 7\ Cl^- + 6\ NO_3^- + 9\ H_2O$$

charges: $7(1^-) \qquad + 6(1^-) = 13^- = 7(1^-) + 6(1^-) = 13^-$

(b) ClO_3^- - reduced (the reduction half-reaction)

ClO_3^- - oxidizing agent

N_2H_4 - oxidized (the oxidizing half-reaction)

N_2H_4 - reducing agent

If your answers are correct, you have completed this chapter. If your answers are incorrect, review Section 16-4 (Examples 16-5, 16-6, and 16-7 and Study Exercise 16-3) in your text.

These problems conclude the chapter on oxidation-reduction equations.

Now take the following quiz to see if you have mastered the material in this chapter.

SAMPLE QUIZ

Quiz #16

1. Calculate the oxidation number for the element indicated in each of the following compounds or ions.

(a) Cl in HClO   (a) _____

(b) P in $P_2O_7{}^{4-}$ (b) _____

2. Balance the following equations by the oxidation number method:

(a) $P + HNO_3 + H_2O \text{ ---> } NO + H_3PO_4$

(b) $FeSO_4 + K_2Cr_2O_7 + H_2SO_4 \text{ ---> }$

$Fe_2(SO_4)_3 + Cr_2(SO_4)_3 + K_2SO_4 + H_2O$

3. In questions 2(a) and (b), what are the oxidizing agents?

2(a) _____ ; 2(b) _____

4. Balance the following equations by the ion electron method:

   (a) $Mn^{2+} + BiO_3^- \longrightarrow MnO_4^- + Bi^{3+}$ in acid solution

   (b) $AsO_3^{3-} + I_2 \longrightarrow AsO_4^{3-} + I^-$ in basic solution

5. In questions 4(a) and (b), what are the reducing agents?

   4(a) _____ ; 4(b) _____

Solutions and Answers for Quiz #16

1. (a) $+1 +$ ox. no. Cl $+ (-2) = 0$
         $+1 +$ ox. no. Cl $-2 = 0$
         ox. no. Cl $-1 = 0$
         ox. no. Cl $= +1$

   (b)   2(ox. no. P) $+ 7(-2) = -4$
         2(ox. no. P) $- 14 = -4$
         2(ox. no. P) $= 10$
         ox. no. P $= +5$

2. (a)
$$\overset{\overset{\displaystyle 3(\text{loss } 5\ e^-)}{\longrightarrow}}{3\ \overset{0}{P} + 5\ H\overset{5+}{N}O_3 + 2\ H_2O \longrightarrow 5\ \overset{2+}{N}O + 3\ H_3\overset{5+}{P}O_4}$$
$$\underset{5(\text{gain } 3\ e^-)}{}$$

   (b)
$$\overset{6(\text{loss } 1\ e^-)}{6\ \overset{2+}{F}eSO_4 + K_2\overset{6+}{C}r_2O_7 + 7\ H_2SO_4 \longrightarrow 3\ \overset{3+}{F}e_2(SO_4)_3 + \overset{3+}{C}r_2(SO_4)_3}$$
$$\underset{1(2 \times \text{gain } 3\ e^-\ \text{per atom})}{}$$
$$+ K_2SO_4 + 7\ H_2O$$

3. (a)   2(a) HNO_3 3. (b)   2(b) $K_2Cr_2O_7$

274

4. (a) $\qquad 2(4\ H_2O + Mn^{2+} \ ---> \ MnO_4^- + 8\ H^+ + 5\ e^-)$

$\qquad 5(6\ H^+ + BiO_3^- + 2\ e^- \ ---> \ Bi^{3+} + 3\ H_2O)$

$\rule{8cm}{0.4pt}$

$\qquad \cancel{8\ H_2O} + 2\ Mn^{2+} + \overset{14}{\cancel{30}}\ H^+ + 5\ BiO_3^- + \cancel{10\ e^-} \ --->$

$\qquad\qquad 2\ MnO_4^- + \cancel{16\ H^+} + \cancel{10\ e^-} + 5\ Bi^{3+} + \overset{7}{\cancel{15}}\ H_2O$

(b) $\qquad 1(2\ OH^- + AsO_3^{3-} \ ---> \ AsO_4^{3-} + H_2O + 2\ e^-)$

$\qquad\qquad 1(2\ e^- + I_2 \ ---> \ 2\ I^-)$

$\rule{8cm}{0.4pt}$

$\qquad 2\ OH^- + AsO_3^{3-} + \cancel{2\ e^-} + I_2 \ ---> \ AsO_4^{3-} + H_2O + 2\ I^- + \cancel{2\ e^-}$

5. (a) 4(a) Mn^{2+} $\qquad\qquad\qquad$ 5. (b) 4(b) AsO_3^{3-}

275

CHAPTER 17

REACTION RATES AND CHEMICAL EQUILIBRIA

In this chapter we will consider reaction rates and chemical equilibria. Reaction rate is the rate or speed at which the products are formed or the reactants consumed. We mentioned equilibria previously in the discussion of vapor pressure, melting or freezing, and saturated solutions. In this chapter we will apply equilibria to chemical reactions. The reactants react to form products and the products react to form reactants, until a state of dynamic equilibria is reached. Reaction rates and chemical equilibria are related because in the course of a reaction the rate of the forward reaction decreases and the rate of the reverse reaction increases until equilibrium is reached.

SELECTED TOPICS

1. The law of mass action states that for a general reaction

$$aA + bB \longrightarrow cC + dD, \qquad (17\text{-}1)$$

the rate of reaction can be defined as

$$k[A]^x[B]^y ; \qquad (17\text{-}2)$$

where [A] and [B] are the concentrations of A and B in moles per liter, respectively; k is the specific rate constant and is determined by experimentation; x and y are usually whole numbers, but maybe fractional numbers or negative numbers or zero, as determined by experimentation.

The rate of a given chemical reaction depends upon (1) the nature of the reactants, (2) the concentration of reactants, (3) the temperature, and (4) the presence of a catalyst. Reactants must collide with sufficient energy before they can react. This energy is expressed as the activation energy. Activation energy is the energy barrier that must be overcome before molecules can react. As the concentrations of the reactants increase, the rate of the reaction also increases. The numerical effect on the rate depends on the value of x and y in the rate equation. Consider the following reaction:

$$2\ H_2(g) + 2\ NO(g) \longrightarrow 2\ H_2O(g) + N_2(g) \qquad (17\text{-}3)$$

with the rate of reaction = $k[H_2][NO]^2$. If the concentration of H_2 is doubled, then the reaction proceeds at 2 times the initial rate [$H_2 = (1 \times 2)^1 = 2$ times the rate]. If the concentration of NO is doubled, then the reaction proceeds at 4 times the initial rate [$NO = (1 \times 2)^2 = 4$ times the rate]. See Section 17-1 in your text.

2. A general equation for a reversible reaction is as follows:

$$A + B \xrightleftharpoons{} C + D \qquad (17\text{-}4)$$

The reaction of A + B ---> C + D is the forward reaction and C + D ---> A + B is the reverse reaction.

At equilibrium, <u>the rate of the forward reaction will be equal to the rate of the reverse reaction</u>. This is a dynamic equilibrium because two opposing reactions are continuously taking place.

For a very general reaction

$$aA + bB \xrightleftharpoons{} cC + dD \qquad (17\text{-}5)$$

where A, B, C, and D represent different molecular species and a, b, c, and d are coefficients in the balanced equation, we can obtain the following expression for the value of K, an equilibrium constant.

$$K = \frac{[C]^c [D]^d}{[A]^a [B]^b} \qquad (17\text{-}6)$$

K is determined experimentally and varies with temperature.

In the equilibrium expression for an equation that contains a solid, we do not consider the solid in the expression, because at constant temperature the concentration in moles per liter of the solid in the solid phase is constant and will not change. This value for the solid is included in K.

For a series of reactions of the same type the larger the value of K, the greater the concentration of the products.

For example, K for acetic acid ($HC_2H_3O_2$) is 1.76×10^{-5} (mol/L) at $25^\circ C$; K for hydrofluoric acid (HF) is 6.80×10^{-4} (mol/L) at $25^\circ C$. Therefore, because the K value for hydrofluoric acid is larger (smaller negative exponent) than it is for acetic acid, more ions are produced in a given volume of solution per mole of hydrofluoric acid than per mole of acetic acid; hence, hydrofluoric acid is a stronger acid than acetic acid. See Section 17-2 in your text.

3. <u>Le Châtelier's Principle</u> states that if an equilibrium system is subject to a change in conditions of concentration, temperature, or pressure, the composition of the reaction mixture will change in such a way as to try to restore the original conditions. Consider the following equation:

$$A(g) + B(g) \xrightleftharpoons{} C(g) + D(g) + heat \qquad (17\text{-}7)$$

Increasing the concentration of A or B will shift the equilibrium to the right. Increasing the concentration of C or D will shift the equilibrium to the left. Removing A or B will shift the equilibrium to the left and removing C or D will shift the equilibrium to the right. The equilibrium constant, K, has a specific value at a specific temperature. If the temperature changes, the value for K changes. In Equation 17-7, the reaction is exothermic, so an increase in temperature causes K to decrease. Le Châtelier's Principle predicts that an increase in temperature will shift the equilibrium to the left, because the heat acts as one of the products. For an endothermic reaction, the opposite would occur; that is, temperature increase would cause K to increase, shifting the equilibrium to the right.

In Equation 17-7, if A, B, C, and D are gases, an increase or decrease in pressure would have no effect on the equilibrium, because two volumes of reactants react to form two volumes of products. For the reaction

$$N_2(g) + 3 H_2(g) \xrightleftharpoons{} 2 NH_3(g) \tag{17-8}$$

a pressure increase would shift the equilibrium to the right, because four volumes of reactants form two volumes of products. Le Châtelier's Principle predicts that increasing the pressure on a system at equilibrium will shift the equilibrium in that direction which will decrease the volume. A catalyst does not affect the position of the equilibrium, because it increases both the forward and reverse reactions to the same degree. See Section 17-3 in your text.

4. Weak electrolytes are ionized only a few percent. Solutions of weak electrolytes contain both the nonionized substance and the ions resulting from the ionization of the weak electrolytes. For weak electrolytes involving acids, the equilibrium constant, K, is called K_a. For weak electrolytes involving bases, the equilibrium constant, K, is called K_b. Problems involving weak electrolytes are of two types: (1) solving for the equilibrium constant, K_a or K_b, given the concentration of the nonionized acid or base in molarity (mol/L) and the percent ionization, and (2) solving for the hydrogen ion or hydroxide ion concentration in molarity (mol/L) and the percent ionization, given the concentration of the nonionized acid or base in molarity (mol/L) and the equilibrium constant (K_a or K_b). Examples 17-3 to 17-5 in your text give examples of these types of problems. See Section 17-4 in your text.

5. A buffer solution is a solution which resists a rapid change in pH. Buffers contain (1) a weak acid and (2) a conjugate base (salt of the weak acid). If acid is added

278

to the buffer solution, the conjugate base will consume the acid to form the weak acid and a salt. If base is added to the buffer solution, the weak acid will consume the base to form the salt of the weak acid and water. The following are examples of buffer solutions, their weak acids, and conjugate base.

Buffer Solution	Weak Acid	Conjugate Base
$HC_2H_3O_2$ and $NaC_2H_3O_2$	$HC_2H_3O_2$	$C_2H_3O_2^-$
NaH_2PO_4 and Na_2HPO_4	$H_2PO_4^-$	HPO_4^{2-}
H_2CO_3 and $NaHCO_3$	H_2CO_3	HCO_3^-

Buffers are very important in the body in that they keep the pH of the blood between 7.3 and 7.5. These blood buffers are:

1. Carbonic acid (H_2CO_3) and hydrogen carbonate or bicarbonate (HCO_3^-)

2. Dihydrogen phosphate $(H_2PO_4^-)$ and hydrogen phosphate $HPO_4^{2-})$

3. certain proteins

Example 17-6 in your text gives more examples of buffers, their weak acids, and conjugate bases. See Section 17-5 in your text.

6. Solubility product equilibria involve a slightly soluble electrolyte in equilibrium with its ions in solution. At equilibrium, the solution is saturated. Consider a slightly soluble electrolyte of the formula A_3B_2 added to pure water. The A_3B_2 will dissolve until a saturated solution is obtained. We express the equation for this equilibrium as follows:

$$A_3B_2\,(\underline{s}) \;\rightleftharpoons\; 3\,A^{2+}\,(\underline{aq}) + 2\,B^{3-}\,(\underline{aq}) \qquad\qquad (17\text{-}9)$$

We could express the equilibrium constant for the equation as:

$$K = \frac{[A^{2+}]^3\,[B^{3-}]^2}{[A_3B_2]}$$

Since A_3B_2 is in the solid state and its concentration in the solid phase is constant, A_3B_2 is incorporated into the constant K and the expression becomes

$$K_{sp} = [A^{2+}]^3 [B^{3-}]^2 \qquad (17\text{-}10)$$

The constant, K_{sp}, is called the solubility product constant. Equation 17-10 defines the solubility product constant in which $[A^{2+}]$ and $[B^{3-}]$ represent the concentration of these ions in molarity (mol/L) in a solution at equilbirium with solid A_3B_2. Therefore, the coefficients of the

ions in a slightly soluble electrolyte in equilibrium with its ions become exponents for the concentrations of the ions in solution in the expression for the solubility product constant, K_{sp}.

The larger the value of the K_{sp} of an electrolyte of the

same type, the more soluble the electrolyte is in water. For example, the solubility product constant for barium

carbonate ($BaCO_3$) is 8.10×10^{-9} (mol^2/L^2) at 25°C; that of

barium sulfate ($BaSO_4$) is 1.08×10^{-10} (mol^2/L^2) at 25°C.

Both are AB-type electrolytes. Therefore, because the K_{sp}

for barium carbonate is larger (smaller negative exponent) than the K_{sp} for barium sulfate, the solubility of barium

carbonate in water is greater than is that of barium sulfate.

Problems involving solubility product equilibria are of two types: (1) solving for the solubility product constant, K_{sp}, given the solubility of the slightly soluble electro-

lyte in water, and (2) solving for the concentration in molarity and grams per liter of the slightly soluble electrolyte given the solubility product constant, K_{sp}, for the

electrolyte. Examples 17-7 to 17-9 in your text give examples of these types of problems. See Section 17-6 in your text.

PROBLEM 17-1

Given the following chemical equation:

$$O_2(g) + 2\ NO(g) \longrightarrow 2\ NO_2(g)$$

with the reaction rate = $k[O_2][NO]^2$, calculate the effect on the reaction rate if:

(a) the concentration of O_2 is doubled from 0.0100 M to 0.0200 M

_____

(b) the concentration of NO is doubled from 0.0100 M to 0.0200 M

_____

(c) the concentration of NO is tripled from 0.0100 M to 0.0300 M

_____

(d) the concentration of NO is increased by 1.50 times from 0.0100 M to 0.0150 M

_____

ANSWERS AND SOLUTION TO PROBLEM 17-1

(a) 2 times   At 0.0100 M   At 0.0200 M
               rate = $k[0.0100\ M][NO]^2$ rate = $k[0.0200\ M][NO]^2$

               The rate would be two times (0.0200/0.0100) the
               rate with 0.0100 M.

(b) 4 times   At 0.0100 M   At 0.0200 M
               rate = $k[O_2][0.0100\ M]^2$ rate = $k[O_2][0.0200\ M]^2$

               = $k[O_2][0.000100\ M^2]$ = $k[O_2][0.0004000\ M^2]$

               The rate would be four times (0.000400/0.000100)
               the rate with 0.0100 M.

(c) 9 times   At 0.0100 M   At 0.0300 M
               rate = $k[O_2][0.0100\ M]^2$ rate = $k[O_2][0.0300\ M]^2$

               = $k[O_2][0.000100\ M^2]$ = $k[O_2][0.0009000\ M^2]$

               The rate would be nine times (0.000900/0.000100)
               the rate with 0.0100 M.

(d) 2.25 times At 0.0100 M   At 0.0150 M
               rate = $k[O_2][0.0100\ M]^2$ rate = $k[O_2][0.0150\ M]^2$

               = $k[O_2][0.000100\ M^2]$ = $k[O_2][0.000225\ M^2]$

               The rate would be 2.25 times (0.000225/0.000100)
               the rate with 0.0100 M.

If your answers are correct, proceed to Problem 17-2. If your answers are incorrect, review Section 17-1, concentration of reactants (Example 17-1 and Study Exercise 17-1) in your text.

Proceed to Problem 17-2.

PROBLEM 17-2

Write the expression for the equilibrium constant for each of the following reactions:

(a) $CH_4(g) + 2 O_2(g) \xrightarrow{\quad} \xleftarrow{\quad} CO_2(g) + 2 H_2O(g)$

_____

(b) $N_2O_4(g) \xrightarrow{\quad} \xleftarrow{\quad} 2 NO_2(g)$

_____

(c) $S(s) + O_2(g) \xrightarrow{\quad} \xleftarrow{\quad} SO_2(g)$

_____

(d) $CuO(s) + H_2(g) \xrightarrow{\quad} \xleftarrow{\quad} Cu(s) + H_2O(g)$

_____

ANSWERS TO PROBLEM 17-2

(a) $K = \dfrac{[CO_2][H_2O]^2}{[CH_4][O_2]^2}$

(b) $K = \dfrac{[NO_2]^2}{[N_2O_4]}$

(c) $K = \dfrac{[SO_2]}{[O_2]}$ The solid (s) is not included in the equilibrium expression.

(d) $K = \dfrac{[H_2O]}{[H_2]}$ The solids (CuO and Cu) are not included in the equilibrium expression.

If your answers are correct, proceed to Problem 17-3. If your answers are incorrect, review Section 17-2, equilibrium constant (Example 17-2 and Study Exercise 17-2) in your text.

Proceed to Problem 17-3.

PROBLEM 17-3

The following are ionization constants for various acids (K_a) at $25^{O}C$. List them in order of decreasing acid strength by their K_a values.

Acid	K_a at $25^{O}C$ (mol/L)
acetic acid	1.76×10^{-5}
benzoic acid	6.30×10^{-5}
chloracetic acid	1.40×10^{-3}
phenylacetic acid	4.88×10^{-5}
p-nitrophenylacetic acid	1.41×10^{-4}

ANSWERS AND SOLUTIONS TO PROBLEM 17-3

If we express the K_a values to the same power of 10, the values are as follows:

Acid	K_a at $25^{O}C$ (mol/L)
acetic acid	1.76×10^{-5}
benzoic acid	6.30×10^{-5}
chloroacetic acid	$14\overline{0} \times 10^{-5}$
phenylacetic acid	4.88×10^{-5}
p-nitrophenylacetic acid	14.1×10^{-5}

Therefore, the acid strengths in decreasing order are as follows:
chloroacetic acid> p-nitrophenylacetic acid> benzoic acid> phenylacetic acid> acetic acid

If your answers are correct, proceed to Problem 17-4. If your answers are incorrect, review Section 17-2, equilibrium constant (Study Exercise 17-3) in your text.

Proceed to Problem 17-4.

PROBLEM 17-4

Consider the following chemical reaction:

$$CO(g) + Cl_2(g) \underset{\longleftarrow}{\overset{\longrightarrow}{}} COCl_2(g)$$

In which direction will the equilibrium be shifted by each of the following changes:

(a) increasing the concentration of CO _____

(b) increasing the concentration of Cl_2 _____

(c) increasing the concentration of $COCl_2$ _____

(d) decreasing the concentration of CO _____

(e) decreasing the concentration of Cl_2 _____

(f) decreasing the concentration of $COCl_2$ _____

ANSWERS TO PROBLEM 17-4

(a) right; (b) right; (c) left; (d) left; (e) left;(f) right

If your answers are correct, proceed to Problem 17-5. If your answers are incorrect, review Section 17-3 (Study Exercise 17-4) in your text.

Proceed to Problem 17-5.

PROBLEM 17-5

Predict the effect on the equilibrium of the following chemical reactions when (1) the temperature in increased, (2) the temperature is decreased, (3) the pressure is increased, and (4) the pressure is decreased:

(a) $N_2(g) + O_2(g) \underset{\longleftarrow}{\overset{\longrightarrow}{}} 2 NO(g) - 181$ kJ

(1) _____ ; (2) _____ ; (3) _____ ; (4) _____

(b) $2 SO_2(g) + O_2(g) \underset{\longleftarrow}{\overset{\longrightarrow}{}} 2 SO_3(g) + 47.0$ kcal

(1) _____ ; (2) _____ ; (3) _____ ; (4) _____

ANSWERS TO PROBLEM 17-5

(a) 181 kJ + N2(g) + O_2(g) $\rightleftharpoons$ 2 NO(g)

     (1) right; (2) left; (3) no effect; (4) no effect

(b)   2 SO_2(g) + O_2(g) $\rightleftharpoons$ 2 SO_3(g) + 47.0 kcal

     (1) left; (2) right; (3) right; (4) left

If your answers are correct, proceed to Problem 17-6. If your answers are incorrect, re-read Section 17-3 (Study Exercise 17-4) in your text.

Proceed to Problem 17-6.

PROBLEM 17-6

Calculate the ionization constant (K_a) for each of the following weak electrolytes from the percent ionization at the concentrations given:

(a) a 0.00400 M solution of $HC_2H_3O_2$ is 6.45 percent ionized

_____

(b)   a 0.500 M solution of HX is 2.00 percent ionized

_____

ANSWERS AND SOLUTIONS TO PROBLEM 17-6

(a) 1.78×10^{-5} (mol/L)

$$HC_2H_3O_2 (aq) \underset{\longleftarrow}{\overset{\dashrightarrow}{}} H^+ (aq) + C_2H_3O_2^- (aq)$$

$$K_a = \frac{[H^+][C_2H_3O_2^-]}{[HC_2H_3O_2]}$$

To evaluate the K_a, we must determine the concentration of H^+ and $C_2H_3O_2^-$ (at equilibrium) from the data given. The $HC_2H_3O_2$ is 6.45 percent ionized, so the percent nonionized at equilibrium would be 93.55 percent (100.00 - 6.45). We calculate the concentrations of H^+, $C_2H_3O_2^-$, and $HC_2H_3O_2$ in solution as follows:

$[H^+]$ = 0.00400 mol/L x 0.0645 = 0.000258 mol/L

$= 2.58 \times 10^{-4}$ mol/L

(The $[C_2H_3O_2^-]$ is equal to the $[H^+]$, because in the balanced equation 1 mol of $HC_2H_3O_2$ ionizes to form 1 mol of H^+ and 1 mol $C_2H_3O_2^-$; therefore 2.58×10^{-4} mol of $HC_2H_3O_2$ would ionize to form 2.58×10^{-4} mol of H^+ and 2.58×10^{-4} mol $C_2H_3O_2^-$).

$[C_2H_3O_2^-]$ = 2.58×10^{-4} mol/L

$[HC_2H_3O_2]$ = 0.00400 mol/L x 0.9355 = 0.00374 mol/L

$= 3.74 \times 10^{-3}$ mol/L at equilibrium

[We can also obtain $[HC_2H_3O_2]$ by subtracting the mol/L of ionized $HC_2H_3O_2$ from the mol/L of the initial $HC_2H_3O_2$, as 0.00400 mol/L (initially) - 0.000258 mol/L (ionized, from the mole relationship in the balanced equation) = 0.00374 mol/L = 3.74×10^{-3} mol/L].

Calculate the value of K_a as follows:

$$K_a = \frac{[H^+][C_2H_3O_2^-]}{[HC_2H_3O_2]}$$

$$= \frac{[2.58 \times 10^{-4} \text{ mol/L}][2.58 \times 10^{-4} \text{ mol/L}]}{[3.74 \times 10^{-3} \text{ mol/L}]}$$

$$= 1.78 \times 10^{-5} \text{ (mol/L)}$$

(b) 2.04×10^{-4} (mol/L)

$$HX\text{(aq)} \quad \overset{\text{---}\rightarrow}{\underset{\leftarrow\text{---}}{}} \quad H^+\text{(aq)} + X^-\text{(aq)}$$

$$K_a = \frac{[H^+][X^-]}{[HX]}$$

Determine the concentrations of H^+ and X^- as follows:

$[H^+] = 0.500$ mol/L $\times 0.0200 = 0.0100$ mol/L

$[X^-] = 0.0100$ mol/L

The HX is 2.00 percent ionized, so the percent non-ionized at equilibrium would be 98.00 percent (100.00 - 2.00). Calculate the concentration of HX at equilibrium as follows:

$[HX] = 0.500$ mol/L $\times 0.9800 = 0.490$ mol/L or 0.500 mol/L (initially) - 0.0100 mol/L (ionized, from the mole relationship in the balanced equation) = 0.490 mol/L

$$K_a = \frac{[H^+][X^-]}{[HX]} = \frac{[0.0100 \text{ mol/L}][0.0100 \text{ mol/L}]}{[0.490 \text{ mol/L}]}$$

$$= \frac{[1.00 \times 10^{-2}][1.00 \times 10^{-2}] (\text{mol}^2/L^2)}{[4.90 \times 10^{-1}] (\text{mol/L})}$$

$$= 2.04 \times 10^{-4} \text{ (mol/L)}$$

If your answers are correct, proceed to Problem 17-7. If your answers are incorrect, review Section 17-4 (Example 17-3 and Study Exercise 17-5) in your text.

Proceed to Problem 17-7.

PROBLEM 17-7

From the ionization constants (K_a or K_b) for each of the following weak electrolytes, calculate the hydrogen ion concentration (for acids) or the hydroxide ion concentration (for bases) in moles per liter and the percent ionization of the weak electrolytes in each of the following solutions:

(a)   0.100 $\underline{M}$ HX. The K_a for HX = 3.00 x 10^{-8} (mol/L)

_____

(b)   0.250 $\underline{M}$ MOH. The K_b for MOH = 2.00 x 10^{-6} (mol/L)

_____

ANSWERS AND SOLUTIONS TO PROBLEM 17-7

(a)   5.48 x 10^{-5} mol/L, 0.0548 %

$$HX\,(\underline{aq}) \; \underset{\longleftarrow ---}{---\longrightarrow} \; H^+\,(\underline{aq}) + X^-\,(\underline{aq})$$

$$K_a = \frac{[H^+]\,[X^-]}{[HX]}$$

In the balanced equation, 1 mol of HX on ionizing

forms 1 mol of H^+ and 1 mol of X^-. Therefore, we

let x = [H^+] in mol/L; x mol/L also equals X^-. The concentration of the nonionized HX remaining at equilibrium is HX = 0.100 mol/L - x mol/L. Substituting

the values for [H^+], [X^-], and [HX] into the equilibrium expression, and dropping the units of x for simplicity, we have:

$$3.00 \times 10^{-8} \text{ (mol/L)} = \frac{[x][x]}{[0.100 \text{ mol/L} - x]}$$

$$3.00 \times 10^{-8} \text{ (mol/L)} = \frac{[x]^2}{[0.100 \text{ mol/L} - x]}$$

Assuming that x is very small in relation to the original concentration of HX (0.100 $\underline{M}$), we drop the x in the denominator and the value of HX is 0.100 mol/L.

$$3.00 \times 10^{-8} \text{ (mol/L)} = \frac{[x]^2}{0.100 \text{ mol/L}}$$

$$x^2 = 3.00 \times 10^{-8} \times 0.100 \ \frac{mol^2}{L^2} = 0.300 \times 10^{-8} \ \frac{mol^2}{L^2}$$

$$x = \sqrt{0.300 \times 10^{-8} \ \frac{mol^2}{L^2}} = 0.548 \times 10^{-4} \text{ mol/L}$$

$$= 5.48 \times 10^{-5} \text{ mol/L}$$

Obtain the square root of 0.300 from your calculator.

To obtain the square root of 10^{-8}, divide the exponent by 2 to give 10^{-4}. You may need to adjust the power of 10 by exponential notation to give an even-numbered exponent. [Check your assumption above:

5.48×10^{-5} mol/L is quite small (~ 0.05 %) with respect to 0.100 mol/L, so our assumption was reasonable.] Calculate the percent ionization at follows:

$$\frac{\text{mol/L of acid ionized}}{\text{mol/L of acid initially}} \times 100$$

The mole per liter of acid ionized is the same as the mole per liter of hydrogen ions, because in the balanced equation the mole relationship is 1:1. Hence,

$$\frac{5.48 \times 10^{-5} \ \text{mol/L}}{0.100 \ \text{mol/L}} \times 100 = \frac{5.48 \times 10^{-5}}{1.00 \times 10^{-1}} \times 10^2 = 0.0548 \ \%$$

(b) 7.07×10^{-4} mol/L, 0.283 %

$$MOH(\underline{aq}) \ \underline{\overline{----}} \ M^+(\underline{aq}) + OH^-(\underline{aq})$$

$$K_{\underline{b}} = \frac{[M^+][OH^-]}{[MOH]}$$

Let x = [OH⁻] in mol/L; therefore, x in mol/L also

equals [M⁺], because in the balanced equation 1 mol

MOH ionizes to form 1 mol M^+ and 1 mol OH⁻. The concentration of the nonionized MOH at equilibrium is [MOH] = 0.250 mol/L - x mol/L, again from the mole relationship in the balanced equation being 1:1. Substituting these values into the equilibrium expression and dropping the units of x, we obtain

$$2.00 \times 10^{-6} \ (mol/L) = \frac{[x][x]}{[0.250 \ mol/L - x]}$$

$$2.00 \times 10^{-6} \ (mol/L) = \frac{[x]^2}{[0.250 \ mol/L - x]}$$

Neglecting the x in 0.250 mol/L - x and further solving the equation, we have

$$2.00 \times 10^{-6} \ (mol/L) = \frac{[x]^2}{0.250 \ mol/L}$$

$$x^2 = 0.250 \ mol/L \times 2.00 \times 10^{-6} \ mol/L$$

$$x^2 = 0.500 \times 10^{-6} \ mol^2/L^2$$

$$x = \sqrt{0.500 \times 10^{-6} \ \frac{mol^2}{L^2}} = 0.707 \times 10^{-3} \ mol/L$$

$$= 7.07 \times 10^{-4} \ mol/L$$

[Check your assumption above: 7.07×10^{-4} mol/L is very small (~ 0.3 %) with respect to 0.250 mol/L, so our assumption is reasonable.] Calculate the percent ionization as follows:

$$\frac{7.07 \times 10^{-4} \ \cancel{mol/L}}{0.250 \ \cancel{mol/L}} \times 100 = \frac{7.07 \times 10^{-4}}{2.50 \times 10^{-1}} \times 10^2 = 0.283 \ \%$$

If your answers are correct, proceed to Problem 17-8. If your answers are incorrect, review Section 17-4 (Example 17-4 and 17-5 and Study Exercise 17-6) in your text.

Proceed to Problem 17-8.

PROBLEM 17-8

Given the following buffer solutions, identify the weak acid and conjugate base and write the equation for the equilibrium involving these species. In addition, write the reactions that would serve to neutralize some added hydrochloric acid (HCl)

and sodium hydroxide (NaOH).

(a)   a solution of benzoic acid ($HC_7H_5O_2$) and sodium benzoate ($NaC_7H_5O_2$)

(b)   a solution of propionic acid ($HC_3H_5O_2$) and sodium propionate ($NaC_3H_5O_2$)

ANSWERS AND SOLUTIONS TO PROBLEM 17-8.

(a)   The weak acid is benzoic acid ($HC_7H_5O_2$) and the conjugate base is benzoate ion ($C_7H_5O_2^-$). The equilibrium is

$$HC_7H_5O_2 (aq) \rightleftharpoons H^+ (aq) + C_7H_5O_2^- (aq)$$

Added hydrochloric acid (a strong acid, so H^+ and Cl^-) would be consumed by the reaction with benzoate ion ($C_7H_5O_2^-$).

$$H^+ (aq) + C_7H_5O_2^- (aq) \longrightarrow HC_7H_5O_2 (aq)$$

Added sodium hydroxide (a strong base, so Na^+ and OH^-) would be consumed by the reaction with benzoic acid ($HC_7H_5O_2$).

$$OH^- (aq) + HC_7H_5O_2 (aq) \longrightarrow C_7H_5O_2^- (aq) + H_2O (\ell)$$

(b)   The weak acid is propionic acid ($HC_3H_5O_2$) and the conjugate base is propionate ion ($C_3H_5O_2^-$). The equilibrium is

$$HC_3H_5O_2 (aq) \rightleftharpoons H^+ (aq) + C_3H_5O_2^- (aq)$$

Added hydrochloric acid (a strong acid, so H^+ and Cl^-) would be consumed by the reaction with propionate ion $(C_3H_5O_2{}^-)$.

$$H^+(\underline{aq}) + C_3H_5O_2{}^-(\underline{aq}) \longrightarrow HC_3H_5O_2(\underline{aq})$$

Added sodium hydroxide (a strong base, so Na^+ and OH^-) would be consumed by the reactions with propionic acid $(HC_3H_5O_2)$.

$$OH^-(\underline{aq}) + HC_3H_5O_2(\underline{aq}) \longrightarrow C_3H_5O_2{}^-(\underline{aq}) + H_2O(\ell)$$

If your answers are correct, proceed to Problem 17-9. If your answers are incorrect, review Section 17-5 (Example 17-6 and Study Exercise 17-7) in your text.

PROBLEM 17-9

The following are solubility product constants (K_{sp}) for various slightly soluble electrolytes in water at $18^\circ C$. List them in order of decreasing solubility in water.

Slightly Soluble Electrolyte	K_{sp} at $18^\circ C$ (mol^2/L^2)
copper(I) chloride	1.0×10^{-6}
copper(I) iodide	5.0×10^{-12}
copper(I) bromide	4.1×10^{-8}

ANSWERS AND SOLUTIONS TO PROBLEM 17-9

If we express the K_{sp} values to the same power of 10, the values are:

Slightly Soluble Electrolyte	K_{sp} at $18^\circ C$ (mol^2/L^2)
copper(I) chloride	1.0×10^{-6}
copper(I) iodide	0.0000050×10^{-6}
copper(I) bromide	0.041×10^{-6}

Therefore, the decreasing solubility in water is:
    copper(I) chloride> copper(I) bromide > copper(I) iodide

If your answers are correct, proceed to Problem 17-10. If your answers are incorrect, review Section 17-6 (Study Exercise 17-8) in your text.

Proceed to Problem 17-10.

PROBLEM 17-10

Write the expression for the solubility product constant (K_{sp}) for each of the following slightly soluble electrolytes.

(a) $FePO_4$

_____

(b) FeC_2O_4

_____

(c) $Al(OH)_3$

_____

(d) $Mn_3(PO_4)_2$

_____

ANSWERS TO PROBLEM 17-10

(a) $FePO_4 (\underline{s}) \underset{\longleftarrow}{\longrightarrow} Fe^{3+} (\underline{aq}) + PO_4^{3-} (\underline{aq})$

$K_{sp} = [Fe^{3+}][PO_4^{3-}]$

(b) $FeC_2O_4 (\underline{s}) \underset{\longleftarrow}{\longrightarrow} Fe^{2+} (\underline{aq}) + C_2O_4^{2-} (\underline{aq})$

$K_{sp} = [Fe^{2+}][C_2O_4^{2-}]$

(c) $Al(OH)_3 (\underline{s}) \underset{\longleftarrow}{\longrightarrow} Al^{3+} (\underline{aq}) + 3 OH^- (\underline{aq})$

$K_{sp} = [Al^{3+}][OH^-]^3$

(d) $Mn_3(PO_4)_2 (\underline{s}) \underset{\longleftarrow}{\longrightarrow} 3 Mn^{2+} + 2 PO_4^{3-}$

$K_{sp} = [Mn^{2+}]^3[PO_4^{3-}]^2$

If your answers are correct, proceed to Problem 17-11. If your answers are incorrect, review Section 17-6 (Example 17-7 and Study Exercise 17-9) in your text.

Proceed to Problem 17-11.

PROBLEM 17-11

From the solubility of each of the following compounds in pure water at a given temperature, calculate the solubility product constant (K_{sp}) for the compound at that temperature:

(a) MX: 1.3×10^{-4} mol/L at 25°C

_____

(b) lead(II) oxalate; 0.0015 g PbC_2O_4/L at 18°C

_____

ANSWERS AND SOLUTIONS TO PROBLEM 17-11

(a) 1.7×10^{-8} (mol^2/L^2)

$$MX(s) \xrightleftharpoons{\quad} M^+(aq) + X^-(aq)$$

$$K_{sp} = [M^+][X^-]$$

The concentration of M^+ and X^- in the solution is 1.3×10^{-4} mol/L each, because 1 mol MX in solution would produce 1 mol each of M^+ and X^- from the balanced equation. The solubility product constant for MX is,

$$K_{sp} = [M^+][X^-] = [1.3 \times 10^{-4} \text{ mol/L}][1.3 \times 10^{-4} \text{ mol/L}]$$

$$= 1.7 \times 10^{-8} \text{ } (mol^2/L^2)$$

(b) 2.6×10^{-11} (mol^2/L^2)

$$PbC_2O_4 (\underline{s}) \underset{\longleftarrow}{\overset{\longrightarrow}{}} Pb^{2+} (\underline{aq}) + C_2O_4{}^{2-} (\underline{aq})$$

$$K_{\underline{sp}} = [Pb^{2+}][C_2O_4{}^{2-}]$$

The $K_{\underline{sp}}$ expression gives the concentration of the ions in moles per liter; hence, we must express the solubility in moles per liter. The molar mass for PbC_2O_4 is 295.2 g, and the concentration in moles per liter (molarity) is,

$$\frac{0.0015 \ \cancel{g \ PbC_2O_4}}{L} \times \frac{1 \ mol \ PbC_2O_4}{295.2 \ \cancel{g \ PbC_2O_4}} = \frac{1.5 \times 10^{-3}}{2.952 \times 10^2} \frac{mol}{L}$$

$$= 0.51 \times 10^{-5} \ mol/L = 5.1 \times 10^{-6} \ mol/L$$

The concentrations of both Pb^{2+} and $C_2O_4{}^{2-}$ in the solution are 5.1×10^{-6} mol/L each, because 1 mol PbC_2O_4 in solution would produce 1 mol each of Pb^{2+} and $C_2O_4{}^{2-}$ from the balanced equation. The solubility product constant for PbC_2O_4 is,

$$K_{\underline{sp}} = [Pb^{2+}][C_2O_4{}^{2-}] = [5.1 \times 10^{-6} \ mol/L][5.1 \times 10^{-6} \ mol/L]$$

$$= 26 \times 10^{-12} \ (mol^2/L^2) = 2.6 \times 10^{-11} \ (mol^2/L^2)$$

If your answers are correct, proceed to Problem 17-12. If your answers are incorrect, review Section 17-6 (Example 17-8 and Study Exercise 17-10) in your text.

Proceed to Problem 17-12.

PROBLEM 17-12

From the solubility product constant for each of the following salts, calculate (1) the molarity of a saturated solution of the salt at the given temperature, and calculate (2) the solubility of the salt in grams per liter at a given temperature:

(a) MX at 25°C; the K_{sp} for MX $= 3.2 \times 10^{-16}$ (mol^2/L^2),

molar mass $= 285.0$ g

_____

(b) CaC_2O_4 at 25°C; the K_{sp} for $CaC_2O_4 = 2.6 \times 10^{-9}$ (mol^2/L^2).

If the concentration of $[Ca^{2+}]$ and $[C_2O_4{}^{2-}]$ solution each

reaches 9.8×10^{-5} mol/L, will precipitation occur?

_____

ANSWERS AND SOLUTIONS TO PROBLEM 17-12

(a) (1) 1.8×10^{-8} $\underline{M}$; (2) 5.1×10^{-6} g/L

(1) MX($\underline{s}$) $\begin{array}{c}\text{----}\rightarrow\\ \leftarrow\text{----}\end{array}$ M^+(a$\underline{q}$) + X^-(a$\underline{q}$)

$K_{sp} = [M^+][X^-]$

Let x = mol MX/L of saturated solution. The concen-

tration of M^+ and X^- is also x mol/L, because from the balanced equation 1 mol of MX in solution yields

296

1 mol M^+ and 1 mol X^-. Hence,

x mol/L = $[M^+]$ = $[X^-]$

and we express the K_{sp} as follows:

$3.2 \times 10^{-16} \ mol^2/L^2 = [M^+][X^-] = [x][x] = x^2$

$x = \sqrt{3.2 \times 10^{-16} \ \dfrac{mol^2}{L^2}} = 1.8 \times 10^{-8} \ mol/L$

Obtain the square root of 3.2 from your calculator. To obtain the square root of 10^{-16}, divide the exponent by 2 to give 10^{-8}. Therefore, the solution is 1.8×10^{-8} mol MX/L or 1.8×10^{-8} M MX.

(2) The molar mass of MX is 285.0 g; calculate the solubility in grams per liter as

$1.8 \times 10^{-8} \ \dfrac{\cancel{mol}}{L} \times \dfrac{285.0 \ g \ MX}{1 \ \cancel{mol}} = 5.1 \times 10^{-6} \ g \ MX/L$

(b) (1) 5.1×10^{-5} M; (2) 6.5×10^{-3} g/L; (3) yes

(1) $CaC_2O_4 (\underline{s}) \ \underset{\longleftarrow}{\dashrightarrow} \ Ca^{2+}(\underline{aq}) + Ca_2O_4{}^{2-}(\underline{aq})$

$K_{sp} = [Ca^{2+}][C_2O_4{}^{2-}]$

Let x = mol CaC_2O_4/L of saturated solution. The concentration of $[Ca^{2+}]$ and $[C_2O_4{}^{2-}]$ is also x mol/L, because from the balanced equation 1 mol CaC_2O_4 in solution yields 1 mol Ca^{2+} and 1 mol $C_2O_4{}^{2-}$. Hence,

x mol/L = $[Ca^{2+}]$ = $[C_2O_4{}^{2-}]$ and K_{sp} is:

$2.6 \times 10^{-9} \ mol^2/L^2 = [Ca^{2+}][C_2O_4{}^{2-}] = [x][x] = [x]^2$

$x = \sqrt{2.6 \times 10^{-9} \ \dfrac{mol^2}{L^2}} = \sqrt{26 \times 10^{-10} \ \dfrac{mol^2}{L^2}}$

$= 5.1 \times 10^{-5} \ mol/L$

Make the exponent even. Then obtain the square root of 26 from your calculator. To obtain the square root of 10^{-10} divide the exponent by 2 to give 10^{-5}.

The solution is 5.1×10^{-5} mol/L or 5.1×10^{-5} M.

(2) The molar mass of CaC_2O_4 is 128.1 g; calculate the solubility in grams per liter as

$$5.1 \times 10^{-5} \frac{mol}{L} \times \frac{128.1 \text{ g}}{1 \text{ mol}} = 6.5 \times 10^{-3} \text{ g } CaC_2O_4/L$$

(3) $[Ca^{2+}] = [C_2O_4{}^{2-}] = 9.8 \times 10^{-5}$ mol/L

The product of the concentration of the ions raised to their respective powers is $[Ca^{2+}]$ $[C_2O_4{}^{2-}]$ = $[9.8 \times 10^{-5}]$ $[9.8 \times 10^{-5}]$ (mol^2/L^2) = 9.6×10^{-9} (mol^2/L^2). This value of 9.6×10^{-9} (mol^2/L^2) is greater than the value of the K_{sp} $[2.6 \times 10^{-9}$ (mol^2/L^2) so yes, precipitation will occur.

If your answers are correct, you have completed this chapter. If your answers are incorrect, review Section 17-6 (Example 17-9 and Study Exercise 17-11) in your text.

These problems conclude the chapter on reaction rates and chemical equilibria.

Now take the sample quizzes to see if you have mastered the material in this chapter.

SAMPLE QUIZZES

The quizzes divide this chapter into two parts. Quiz 17A (Section 17-1 to 17-3) and Quiz 17B (Sections 17-4 to 17-6).

Quiz #17A (Section 17-1 to 17-3)

1. Write the expression for the equilibrium constant for each of the following reactions.

(a) $BaSO_3$ (s) $\rightleftharpoons$ BaO (s) + SO_2 (g)

(b) $COBr_2$ (g) $\rightleftharpoons$ CO (g) + Br_2 (g)

298

(c) $CaCO_3(s)$ $\underset{\longleftarrow}{\dashrightarrow}$ $CaO(s) + CO_2(g)$

(d) $2\ NO(g) + Cl_2(g)$ $\underset{\longleftarrow}{\dashrightarrow}$ $2\ NOCl(g)$

2.   The following are ionization constants for various bases (K_b) at $25^\circ C$. List them in order of decreasing strength according to their K_b values.

Base (in approximately 0.1 N aqueous solution)	K_b at $25^\circ C$ (mol/L)
pyridine	1.7×10^{-9}
ammonia	1.79×10^{-5}
nicotine	7×10^{-7}
codeine	9×10^{-9}

3.   Consider the following chemical reaction:

   $CO(g) + H_2O(g)$ $\underset{\longleftarrow}{\dashrightarrow}$ $CO_2(g) + H_2(g)$

(a)   Predict the effect on equilibrium when the concentration of $H_2O(g)$ is increased

_____

(b)   Predict the effect on equilibrium when the concentration of $CO_2(g)$ is increased

_____

(c)   Predict the effect on equilibrium when the concentration of $CO(g)$ is decreased

_____

(d)   Predict the effect on equilibrium when the concentration of $H_2(g)$ is decreased

_____

Solutions and Answers for Quiz #17A

1. (a) $K = [SO_2]$; (b) $K = \frac{[CO][Br_2]}{[COBr_2]}$; (c) $K = [CO_2]$

(d) $K = \frac{[NOCl]^2}{[NO]^2[Cl_2]}$

2. ammonia> nicotine> codeine> pyridine

3. (a) right; (b) left; (c) left; (d) right

Quiz #17B (Sections 17-4 to 17-6)

1. Write the expression for the solubility product constant (K_{sp}) for each of the following slightly soluble electrolytes:

(a) PbF_2

(b) $Fe(OH)_3$

2. Calculate the ionization constant (K_a) for HA if 0.300 M solution of HA is 1.75 % ionized.

3. Calculate the hydrogen ion concentration in moles per liter and the percent ionization for a 0.150 M HA solution.

[$K_a = 2.45 \times 10^{-8}$ (mol/L) for HA at 25°C]

4. Calculate the solubility product constant at 25°C for the salt MX, it the solubility of the salt in water at 25°C is 1.4×10^{-4} g/L. (Atomic mass: M = 40.0 amu, X = 90.0 amu)

5. Calculate the solubility of the slightly soluble salt (BC) in grams per liter at 25°C, if the K_{sp} for BC is 8.8×10^{-15} (mol^2/L^2). (Atomic mass: B = 15.0 amu; C = 25.0 amu)

Solutions and Answers for Quiz #17B

1. (a) $K_{sp} = [Pb^{2+}][F^-]^2$

    (b) $K_{sp} = [Fe^{3+}][OH^-]^3$

2. $HA(aq) \xrightleftharpoons{} H^+(aq) + A^-(aq)$

    $K_a = \dfrac{[H^+][A^-]}{[HA]}$

    $[H^+] = [A^-] = 0.300$ mol/L x $0.0175 = 5.25 \times 10^{-3}$ mol/L

    $[HA] = 0.300$ mol/L x $0.9825 = 0.295$ mol/L or

    0.300 mol/L $- 0.00525$ mol/L $= 0.295$ mol/L

    $K_a = \dfrac{[5.25 \times 10^{-3}\ \text{mol/L}][5.25 \times 10^{-3}\ \text{mol/L}]}{[2.95 \times 10^{-1}\ \text{mol/L}]}$

    $= 9.34 \times 10^{-5}$ (mol/L)

3. $HX \xrightarrow[\xleftarrow{\text{- - -}}]{\text{- - ->}} H^+ + X^-$

$K_a = \dfrac{[H^+][X^-]}{[HX]}$

Let $x = [H^+] = [X^-]$

$[HX] = 0.150 \text{ mol/L} - x \text{ mol/L}$

$2.45 \times 10^{-8} \text{ mol/L} = \dfrac{[x][x]}{[0.150 \text{ mol/L} - x]}$

$2.45 \times 10^{-8} \text{ mol/L} = \dfrac{[x]^2}{0.150 \text{ mol/L}}$

$x^2 = 2.45 \times 10^{-8} \times 0.150 \dfrac{mol^2}{L^2} = 0.368 \times 10^{-8} \dfrac{mol^2}{L^2}$

$x = \sqrt{0.368 \times 10^{-8} \dfrac{mol^2}{L^2}} = 0.607 \times 10^{-4} \text{ mol/L}$

$\qquad\qquad\qquad\qquad\qquad = 6.07 \times 10^{-5} \text{ mol/L}$

$\dfrac{6.07 \times 10^{-5} \text{ mol/L}}{0.150 \text{ mol/L}} \times 100 = 0.0405 \%$

4. $MX(\underline{s}) \xrightarrow[\xleftarrow{\text{- - -}}]{\text{- - ->}} M^+(\underline{aq}) + X^-(\underline{aq})$

$K_{sp} = [M^+][X^-]$

The molar mass of MX is 130.0 g. The concentration in moles per liter (molarity) is:

$1.4 \times 10^{-4} \dfrac{\text{g MX}}{L} \times \dfrac{1 \text{ mol}}{130.0 \text{ g MX}} = 0.011 \times 10^{-4} \text{ mol/L}$

$[M^+] = [X^-] = 1.1 \times 10^{-6} \text{ mol/L}$

$K_{sp} = [1.1 \times 10^{-6} \text{ mol/L}][1.1 \times 10^{-6} \text{ mol/L}]$

$\qquad\qquad\qquad\qquad = 1.2 \times 10^{-12} \ (mol^2/L^2)$

5. $BC(\underline{s}) \xrightleftharpoons{\hspace{1cm}} B^+(\underline{aq}) + C^-(\underline{aq})$

$K_{\underline{sp}} = [B^+][C^-]$

Let x mol/L = $[B^+]$ = $[C^-]$

$8.8 \times 10^{-15} \frac{mol^2}{L^2} = [B^+][C^-] = x^2$

$x = \sqrt{8.8 \times 10^{-15} \frac{mol^2}{L^2}} = \sqrt{88 \times 10^{-16} \frac{mol^2}{L^2}}$

$\hspace{6cm} = 9.4 \times 10^{-8} \text{ mol/L}$

$9.4 \times 10^{-8} \frac{mol}{L} \times \frac{40.0 \text{ g}}{1 \text{ mol}} = 380 \times 10^{-8} \text{ g/L} = 3.8 \times 10^{-6} \text{ g/L}$

CHAPTER 18

ORGANIC CHEMISTRY

Organic chemistry is the study of substances containing carbon.
A typical organic compound is common table sugar ($C_{12}H_{22}O_{11}$).
Some useful organic substances probably of interest to you are
textiles derived from natural fibers (cotton, wool) and
synthetic polyamide and polyester fibers (nylon and Dacron, respectively); vitamins (A, B_1, B_2, B_6, B_{12}, C, D, E, K); hor-
mones (estrone, progesterone, testosterone, insulin, cort-
icosterone, epinephrine); and medicinals such as aspirin,
caffeine, antihistamine drugs, and antibiotics (penicillins,
cephalosporins, erythromycin, and tetracyclines).

SELECTED TOPICS

1. Organic compounds fall into two broad cetegories:

    (1) hydrocarbons (organic compounds that contain only the
        elements carbon and hydrogen)
    (2) derivatives of hydrocarbons.

    The hydrocarbons are divided into aliphatic hydrocarbons
    and aromatic hydrocarbons. The aliphatic hydrocarbons are
    further divided into: (1) alkanes, (2) alkenes, and (3)
    alkynes. This division of organic compounds is summarized
    in Figure 18-1 in your text. See Section 18-1.

2. Carbon exists in three basic bonding arrangements: (1)
    carbon atoms bonded to four atoms or groups of atoms where
    carbon has bond angles of 109.5° and forms four single
    bonds to give a tetrahedral arrangement; (2) carbons atoms
    bonded to three atoms or group of atoms where carbon has
    bond angles of 120° and forms a double bond and two single
    bonds to give a planar arrangement; and (3) carbon atoms
    bonded to two atoms or groups of atoms where carbon has
    bond angles of 180° and forms either a triple bond and a
    single bond or two double bonds to give a linear arrange-
    ment. Methane (CH_4) is an example of tetrahedral (1);
    ethylene ($CH_2 = CH_2$) is an example of planar (2); and
    acetylene (H-C $\equiv$ C-H) and carbon dioxide (O $=$ C $=$ O) are
    examples of linear (3).

    There are two general methods for drawing the structure of
    organic molecules. They are: (1) structural formulas and
    (2) condensed structural formulas. Structural formulas
    have previously been considered in Section 6-6 in your text
    as an extension of Lewis structures with a dash (——) to
    denote a pair of electrons shared between atoms. (Lewis
    structures are useful, but do become cumbersome for large
    molecules.) In condensed structural formulas, the hydrogen
    atoms are written collectively next to the carbon atom to
    which they are attached. An example is propane (C_3H_8).

```
   H   H   H
   |   |   |
H—C—C—C—H          CH₃-CH₂-CH₃   or    CH₃CH₂CH₃
   |   |   |
   H   H   H
```

$CH_3-CH_2-CH_3$ or $CH_3CH_2CH_3$

structural formula   condensed structural formula
of propane   of propane

Although propane is drawn as a "straight chain," in reality
it is not. The center carbon has a tetrahedral geometry
and the C   C   C bond angle is approximately 109.5°.
Therefore, the "straight chain" is actually a continuous
chain of carbon atoms.

Carbon atoms can also be arranged in circles or rings as in
cyclopentane (C_5H_{10}) and cyclohexane (C_6H_{12}).

           cyclopentane   cyclohexane

Each corner of the ring represents a $-CH_2-$.

See Section 18-2 in your text.

3. Alkanes (al'kans) are aliphatic hydrocarbons that have the
    general molecular formula C_nH_{2n+2} for open chain compounds.

    They are also called saturated hydrocarbons. The simplest
    alkane is methane (CH_4). Table 18-4 of your text gives the
    alkanes from methane to decane. You must know the names
    and structures of the alkanes in this table.

    Isomers are compounds that have the same molecular formula
    but different structural formulas. One isomer of a given
    molecular formula has different properties than another
    isomer of the same molecular formula. The two isomers of
    C_4H_{10} are:

                              305

$$CH_3-CH_2-CH_2-CH_3 \qquad\qquad CH_3-CH-CH_3$$
$$\qquad\qquad\qquad\qquad\qquad\qquad\qquad\quad |$$
$$\qquad\qquad\qquad\qquad\qquad\qquad\qquad CH_3$$

butane   isobutane
m.p. $-138^{O}C$; b.p. $0^{O}C$ m.p. $-159^{O}C$; b.p. $-12^{O}C$

In drawing isomers from the molecular formulas, follow the various guidelines in Section 18-3 of your text.

Alkanes are named by a systematic method of nomenclature called the IUPAC (International Union of Pure and Applied Chemistry) system. Basically this system uses names composed of two parts: (1) the terminal portion names the longest continuous chain in the molecule, the parent chain; (2) the beginning portion names the substituent groups attached to the parent chain. The substituent groups are frequently alkyl groups. The alkyl groups are derived by removing one hydrogen atom from an alkane and are named generally by replacing the -ane of alkane by -yl. You must learn the trivial names and the formulas of the alkyl groups as summarized in Table 18-5 of your text. As the parent structure, the terminal portion of the alkanes uses the names and formulas of the alkanes given in Table 18-4 of your text. Section 18-3, nomenclature, in your text gives rules and examples for naming alkanes by IUPAC names. The carbons in the chain are numbered starting at the end of the chain that gives the lowest numbers to the group or groups attached to the parent structure. You should study these examples.

In writing the structural formula of an alkane from its name, follow the steps given in Section 18-3, nomenclature of your text. Remember that each carbon atom has four covalent bonds.

Alkanes are relatively inert. They do not react under ordinary conditions with acids, bases, oxidizing agents, or reducing agents. They do react with oxygen to produce carbon dioxide, water, and heat energy which heats homes and powers automobiles. Another reaction of alkanes is halogenation, in which a halogen atom (chlorine or bromine) with sufficient heat or light, replaces a hydrogen atom. This type of reaction is a substitution reaction. Section 18-3, uses and reactions of alkanes, in your text gives a general and specific example of monohalogenation substitution reactions of an alkane.

See Section 18-3 in your text.

4. <u>Alkenes</u> (al'kens) are aliphatic hydrocarbons that have the general molecular formula C_nH_{2n} for open chain compounds.

The simplest alkene is ethylene (C_2H_4). Both the alkenes and the alkynes are called unsaturated hydrocarbons, because they contain fewer than the maximum number of hydrogen atoms in their general molecular formula. The alkenes contain <u>double</u> bonds. The alkanes do not contain double bonds and are called saturated hydrocarbons.

Section 18-4, nomenclature, in your text gives rules for naming alkenes by the IUPAC system. It is important to determine the longest continuous chain of carbons containing the double bond and to use this chain as the parent structure. Number the parent structure to give the lowest possible number to the double bond. You should study the various examples of names of alkenes given in this section of your text.

In writing structural formulas of alkenes from the names, we follow the same steps we took with alkanes except we add the double bond in the correct position in the parent structure. Section 18-4, nomenclature, in your text gives examples.

Alkenes undergo <u>addition</u> <u>reactions</u>. One of the most important addition reactions of alkenes is <u>polymerization</u>. In this process, polymers are formed. <u>Polymers</u> (many parts) are substances made of thousands of smaller molecules (monomers) that have bonded together to form a giant molecule. These polymers are called <u>addition</u> <u>polymers</u> since they are formed from alkene monomers that hook together by breaking one part of the double bond and forming two new single bonds. Examples of addition polymers are polethylene, Teflon, and polyvinyl chloride.

Addition reactions occur across the double bond and break this bond. The alkenes usually react with bromine or chlorine at room temperature to form the di-bromo or di-chloro compound, breaking the double bond. Section 18-4, uses and reactions of alkenes, in your text gives a general and specific example of an addition reaction of an alkene with a halogen (bromine or chlorine).

See Section 18-4 in your text.

5. <u>Alkynes</u> (al'kins) are aliphatic hydrocarbons that have the general molecular formula C_nH_{2n-2} for open chain compounds.

The simplest alkyne is acetylene (C_2H_2). Like the alkenes, the alkynes are called unsaturated hydrocarbons. The alkynes contain a <u>triple</u> bond.

Section 18-6, nomenclature, in your text gives rules for naming alkynes by the IUPAC system. As with alkenes, it is important to determine the longest continuous chain of carbons (containing, in this case, the triple bond) and to use this chain as the parent structure. Number the parent structure to give the lowest number to the triple bond. You should study the various examples of names of alkynes in this section of your text.

In writing structural formulas of alkynes, we follow the same steps we took with alkanes except we add the triple bond in the correct position in the parent structure. Section 18-6 , nomenclature, in your text gives examples.

Alkynes usually react with chlorine or bromine (halogenation) by addition reactions at room temperature. The reactions occur across the triple bond, breaking two of the three bonds. Therefore, two moles of the halogen are used. Section 18-6, uses and reactions of alkynes, gives a general and specific example of an addition reaction of an alkyne with a halogen (bromine or chlorine).

See Section 18-5 in your text.

6. The aromatic hydrocarbons are hydrocarbons that have a ring of carbon atoms (benzene ring) and alternating carbon-carbon double bonds within that ring. Benzene (C_6H_6) is the simplest aromatic hydrocarbon. Benzene is a cyclic molecule. We draw the formula of benzene in several ways.

The double bonds are alternated around the hexagon ring. The two-headed arrow between the two formulas indicates these alternating double bonds. Each corner represents a C-H; the circle represents the alternating double bonds using one formula instead of two. We will use the circle method of representing benzene.

Section 18-6, nomenclature, in your text gives rules for naming aromatic hydrocarbons by the IUPAC system. The following structures show two substituents:

ortho (or 1,2-) meta (or 1,3-) para (or 1,4-)
abbreviated o-   abbreviate m- abbreviated p-

For two substituents and three or more substituents, the parent structure is either benzene or toluene.

toluene

Study the various examples of names of aromatic hydrocarbons in this section.

In writing structures of aromatic compounds from the names, we draw the structure of the parent compound and then attach the various groups. Section 18-6, nomenclature, in your text gives examples.

Aromatic compounds undergo halogenation by substitution. A halogen atom (Cl or Br) group replaces a hydrogen atom from the aromatic ring. Section 18-6, uses and reactions of aromatic compounds, in your text gives examples of these substitution reactions. See Section 18-6 in your text.

Table 18-6 in your text summarizes the hydrocarbons (alkanes, alkenes, alkynes, and aromatics). You should use Table 18-6 as a review.

7. Hydrocarbon derivatives are formed from hydrocarbons by replacing one hydrogen on the hydrocarbon with a functional group. A functional group is an atom or group of atoms (other than hydrogen) that is attached to a hydrocarbon chain and that confers some distinctive chemical or physical properties on the organic compound. The properties of the hydrocarbon derivative depends on the functional group. Examples of hydrocarbon derivatives are alkenes, alkynes, organic halides, alcohols, ethers, aldehydes, ketones, carboxylic acids, esters, amines, and amides. Table 18-7 lists these compounds along with their general formulas and functional groups. You must know the names of these compounds, their general formulas, functional groups, and be able to identify the functional group in a given compound. See Section 18-7 in your text.

309

PROBLEM 18-1

Identify the geometry (tetrahedral, 109.5°; planar, 120°; linear, 180°) at all carbon atoms in each of the following molecules:

(a) C_2H_6, ethane,

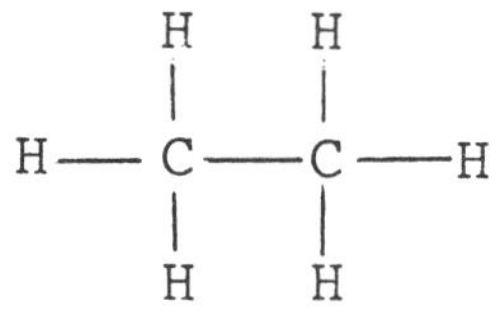

```
      H   H
      |   |
H —— C —— C —— H
      |   |
      H   H
```

(b) C_4H_6, 2-butyne,

```
      H                 H
      |                 |
H —— C —— C≡C —— C —— H
      |                 |
      H                 H
```

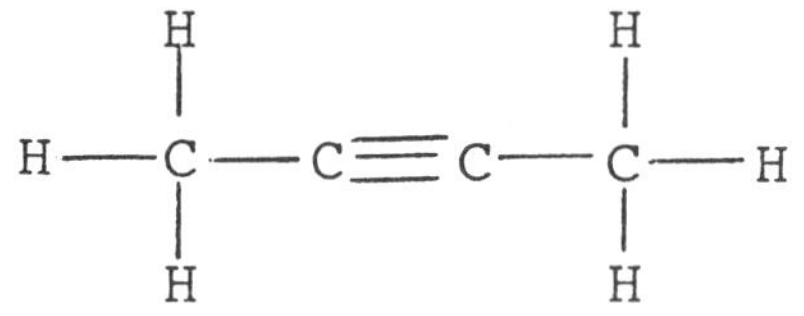

ANSWERS AND SOLUTIONS TO PROBLEM 18-1

(a) tetrahedral, 109.5°

Each carbon has four single bonds, so ethane is tetrahedral The bond angles are 109.5°.

(b) linear, 180°; and tetrahedral, 109.5°

The triply bonded carbon atoms have a triple bond and a single bond, so they are linear. The bond angles are 180°. The other two carbon atoms are tetrahedral because each carbon has four single bonds. The bond angles are 109.5°.

If your answers are correct, proceed to Problem 18-2. If your answers are incorrect, review Section 18-2 (Example 18-1 and Study Exercise 18-1) in your text.

Proceed to Problem 18-2.

PROBLEM 18-2

Write condensed structural formulas for the isomers of the following: (The number in parentheses is the number of isomers for the compound.)

(a) C_5H_{12} (3)

_____   _____

_____

(b) C_6H_{14} (5)

_____   _____

_____   _____

_____

ANSWERS AND SOLUTIONS TO PROBLEM 18-2

(a) C_5H_{12}

> Guideline 1: Continuous chain of carbons.
>
> > (1) C—C—C—C—C
>
> Guideline 2: Remove one carbon atom and place it on another carbon atom so that the new skeleton differs from the previous carbon skeleton.
>
> > (2) 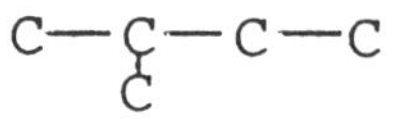 C—C—C—C   C—C—C—C   is the same as
> > | | (1) and 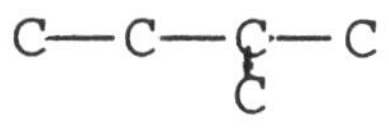
> >   C   C
> >
> >   C—C—C—C   is the same as
> > | (2).
> >   C
>
> Guideline 3: Remove two carbon atoms and place them as single carbon atoms on other carbon atoms in the chain.
>
> > (3) 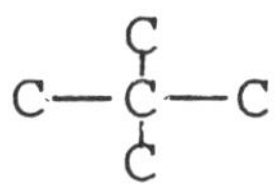   C
> > |
> >   C—C—C
> > |
> >   C
>
> Guideline 4: Place H atoms on the carbon atoms (four bonds to each carbon atom).
>
> > (1) $CH_3-CH_2-CH_2-CH_2-CH_3$
> >
> > (2) 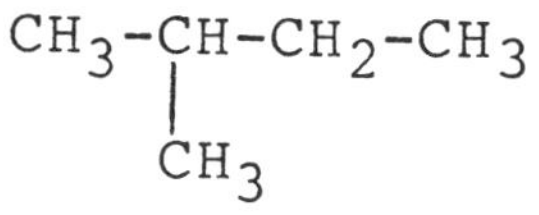 $CH_3-CH-CH_2-CH_3$
> > |
> > CH_3
> >
> > (3) CH_3
> > |
> > CH_3-C-CH_3
> > |
> > CH_3

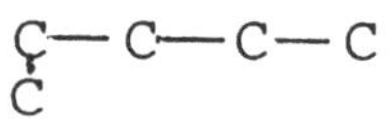

311

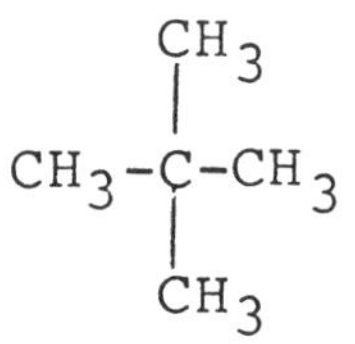

(b) C_6H_{14} (5)

Guideline 1: (1) C—C—C—C—C—C

Guideline 2: (2)
```
    C—C—C—C—C
        |
        C
```

Guideline 2: (3)
```
    C—C—C—C—C
        |
        C
```

Guideline 3: (4)
```
    C—C—C—C
      |   |
      C   C
```

Guideline 3: (5)
```
        C                     C
        |                     |
    C—C—C—C    and    C—C—C—C   are the same.
        |                 |
        C                 C
```

Guideline 4: (1) $CH_3-CH_2-CH_2-CH_2-CH_2-CH_3$

(2)
```
    CH3-CH-CH2-CH2-CH3
       |
       CH3
```

(3)
```
    CH3-CH2-CH-CH2-CH3
            |
            CH3
```

(4)
```
    CH3-CH—CH-CH3
       |    |
       CH3  CH3
```

(5)
```
            CH3
            |
    CH3-CH2-C-CH3
            |
            CH3
```

If your answers are correct, proceed to Problem 18-3. If your answers are incorrect, review Section 18-3, isomers (Example 18-2 and Study Exercise 18-2) in your text.

Proceed to Problem 18-3.

PROBLEM 18-3

Write the IUPAC name for each of the following compounds:

(a) 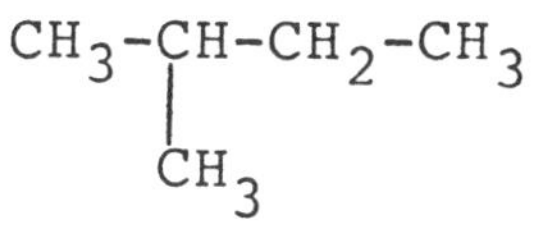 $CH_3-CH-CH_2-CH_3$
            |
           CH_3

_____

(b)  CH_3
         \
          [cyclopentane ring]
                    —CH_3

_____

      CH_3 CH_3 Br
        | | |
(c) 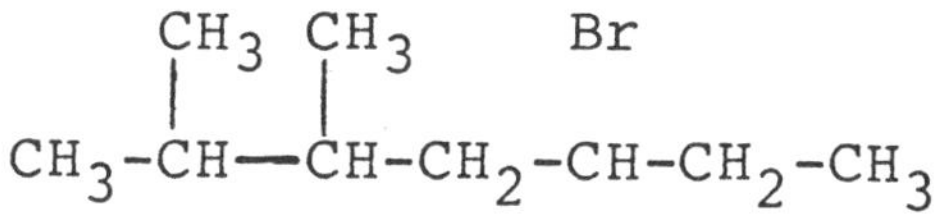 $CH_3-CH—CH-CH_2-CH-CH_2-CH_3$

_____

ANSWERS TO PROBLEM 18-3

(a)   2-methylbutane

(b)   1,3-dimethylcyclopentane

         ([pentagon] is cyclopentane. Give the lowest possible
         numbers to the substituents.)

(c)   5-bromo-2,3-dimethylheptane

         (Give the lowest possible numbers to the alkyl
         substituents. Be sure to include a number for
         each substituent, such as 2,3-dimethyl-.)

If your answers are correct, proceed to Problem 18-4. If your
answers are incorrect, review Section 18-3, nomenclature
(Example 18-3 and Study Exercise 18-3) in your text.

Proceed to Problem 18-4.

PROBLEM 18-4

Write the structural formula for each of the following compounds:

(a)   3,3,4-trimethyloctane

_____

313

(b) 1,1,3-trimethylcyclobutane

_____

(c) 1-bromo-2-chloro-3,4-dimethylhexane

_____

ANSWERS TO PROBLEM 18-4

(a) 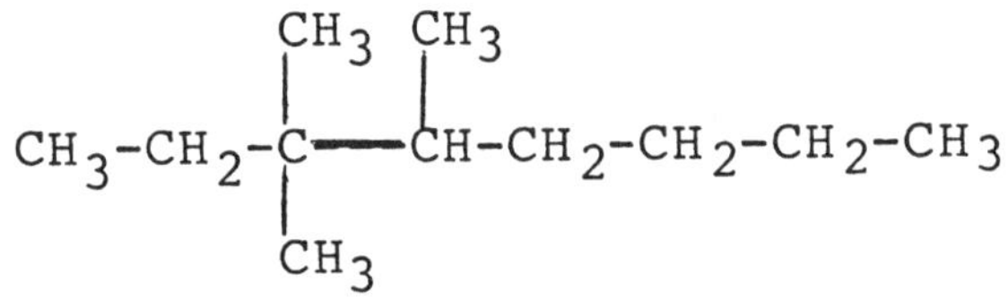 $CH_3-CH_2-\overset{\overset{\displaystyle CH_3}{|}}{\underset{\underset{\displaystyle CH_3}{|}}{C}}$═$\overset{\overset{\displaystyle CH_3}{|}}{CH}-CH_2-CH_2-CH_2-CH_3$

(Octane has 8 carbons.)

(b)

CH_3───$\square$───$\overset{\displaystyle CH_3}{}$───$CH_3$

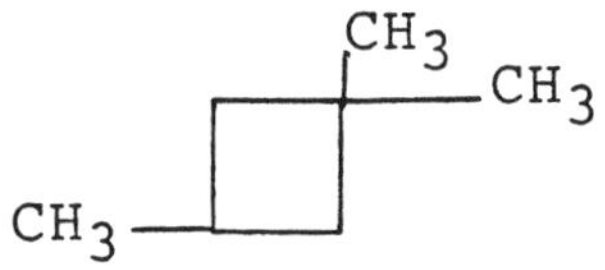

($\square$ is cyclobutane. All positions are equivalent on the ring.)

(c) 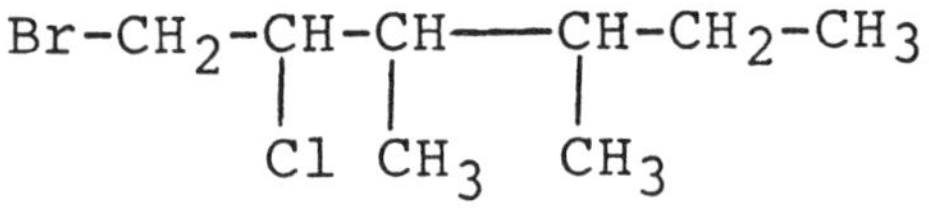 $Br-CH_2-\overset{\overset{\displaystyle }{|}}{\underset{\underset{\displaystyle Cl}{|}}{CH}}-\overset{\overset{\displaystyle }{|}}{\underset{\underset{\displaystyle CH_3}{|}}{CH}}$───$\overset{\overset{\displaystyle }{|}}{\underset{\underset{\displaystyle CH_3}{|}}{CH}}-CH_2-CH_3$

If your answers are correct, proceed to Problem 18-5. If your answers are incorrect, review Section 18-3, nomenclature Example 18-4 and Study Exercise 18-4) in your text.

Proceed to Problem 18-5.

314

PROBLEM 18-5

Complete and balance the reaction equation for monohalogenation in each of the following chemical reactions:

(a) CH_4 + Br_2 $\xrightarrow[\text{light}]{\triangle \text{ or}}$

(b) CH_3-CH_3 + Cl_2 $\xrightarrow[\text{light}]{\triangle \text{ or}}$

ANSWERS TO PROBLEM 18-5

Monohalogenation uses only one mole of the halogen.

(a)
$$\begin{array}{c} H \\ | \\ H-C-H \\ | \\ H \end{array} + Br-Br \xrightarrow[\text{light}]{\triangle \text{ or}} \begin{array}{c} H \\ | \\ H-C-Br \\ | \\ H \end{array} + HBr(g)$$

(b)
$$\begin{array}{c} H\ H \\ |\ | \\ H-C-C-H \\ |\ | \\ H\ H \end{array} + Cl-Cl \xrightarrow[\text{light}]{\triangle \text{ or}} \begin{array}{c} H\ H \\ |\ | \\ H-C-C-Cl \\ |\ | \\ H\ H \end{array} + HCl(g)$$

If your answers are correct, proceed to Problem 18-6. If your answers are incorrect, review Section 18-3, uses and reaction of alkane (Study Exercise 18-5) in your text.

Proceed to Problem 18-6.

PROBLEM 18-6

Write the IUPAC name for each of the following compounds:

(a) $CH_3-CH\!\!=\!\!CH-CH_2-\underset{\underset{CH_3}{|}}{CH}-CH_2-CH_3$

_____

(b)
$$CH_3-\underset{\underset{CH_3}{|}}{\overset{\overset{CH_3}{|}}{C}}-CH\!\!=\!\!CH-\underset{\overset{|}{CH_3}}{CH}-CH_2-CH_3$$

_____

(c)

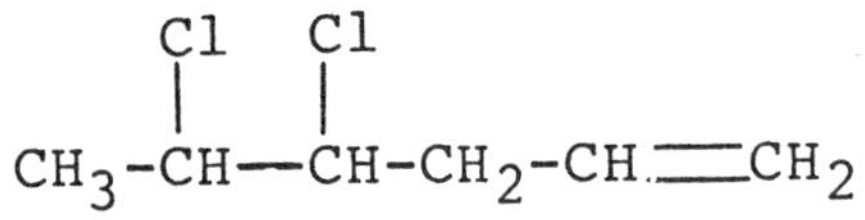

$$CH_3-\overset{\overset{\displaystyle Cl}{|}}{CH}-\overset{\overset{\displaystyle Cl}{|}}{CH}-CH_2-CH\!\!=\!\!CH_2$$

ANSWERS TO PROBLEM 18-6

(a) 5-methyl-2-heptene (Give the lowest possible number to the double bond.)

(b) 2,2,5-trimethyl-3-heptene

(c) 4,5-dichloro-1-hexene   (Give the lowest possible number to the double bond by numbering from right to left in the chain.)

If your answers are correct, proceed to Problem 18-7. If your answers are incorrect, review Section 18-4, nomenclature (Example 18-5 and Study Exercise 18-6) in your text.

Proceed to Problem 18-7.

PROBLEM 18-7

Write the structural formula for each of the following compounds:

(a) 3-bromo-2-methyl-1-pentene

(b) 4-ethyl-5-methyl-3-heptene

(c) 3-bromocyclohexene

ANSWERS TO PROBLEM 18-7

(a) $CH_2\!=\!C\!-\!CH\!-\!CH_2\!-\!CH_3$
            | |
           CH_3 Br

(b) $CH_3\!-\!CH_2\!-\!CH\!=\!C\!-\!CH\!-\!CH_2\!-\!CH_3$
                        | |
                       CH_2 CH_3
                        |
                       CH_3

(c)   (The double bond can be in any position on
                             the ring. The first carbon in the double
                             bond is the number 1 position.)

If your answers are correct, proceed to Problem 18-8. If your
answers are incorrect, review Section 18-4, nomenclature
(Example 18-6 and Study Exercise 18-7) in your text.

Proceed to Problem 18-8.

PROBLEM 18-8

Complete and balance the reaction equation for each of the following chemical reactions:

(a) $CH_3\!-\!CH\!-\!CH\!=\!CH_2$ + Br_2 ---->
            |
           CH_3 in CCl_4

(b) $CH_3\!-\!CH\!-\!C\!=\!CH_2$ + Br_2 ---->
            | |
           CH_3 CH_3 in CCl_4

ANSWERS TO PROBLEM 18-8

(a) $CH_3\!-\!CH\!-\!CH\!=\!CH_2$ + Br-Br ----> $CH_3\!-\!CH\!-\!CH\!-\!CH_2$
            | in CCl_4 | | |
           CH_3 CH_3 Br   Br

                             (Only one bond breaks of the
                             double bond. None of the C-C
                             single bonds break.)

                                                          Br Br
                                                          | |
(b) $CH_3\!-\!CH\!-\!C\!=\!CH_2$ + Br-Br ----> $CH_3\!-\!CH\!-\!C\!-\!CH_2$
            | | in CCl_4 | |
           CH_3 CH_3 CH_3 CH_3

317

If your answers are correct, proceed to Problem 18-9. If your answers are incorrect, review Section 18-4 , uses and reactions of alkenes (Study Exercise 18-8) in your text.

Proceed to Problem 18-9.

PROBLEM 18-9

Write the IUPAC name for each of the following compounds:

(a) $CH_3-C{\equiv}C-CH_2-CH_2-Br$

_____

(b) 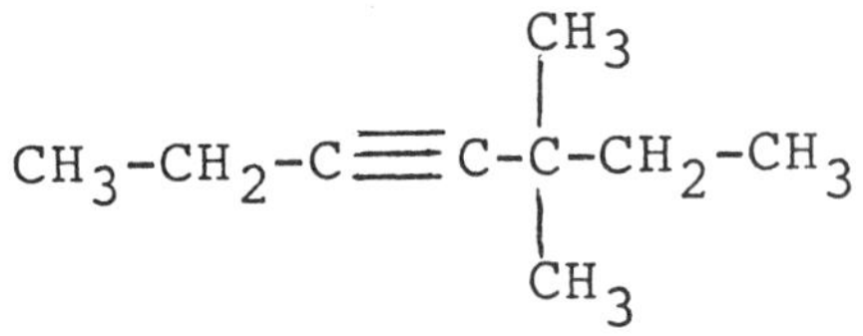 $CH_3-CH_2-C{\equiv}C-\overset{\overset{\displaystyle CH_3}{|}}{\underset{\underset{\displaystyle CH_3}{|}}{C}}-CH_2-CH_3$

_____

(c) $CH_3-CH_2-\overset{\overset{\displaystyle CH_3}{|}}{\underset{\underset{\displaystyle CH_3}{|}}{C}}-C{\equiv}C-CH_2-\overset{}{\underset{\underset{\displaystyle Cl}{|}}{C}}H-Cl$

_____

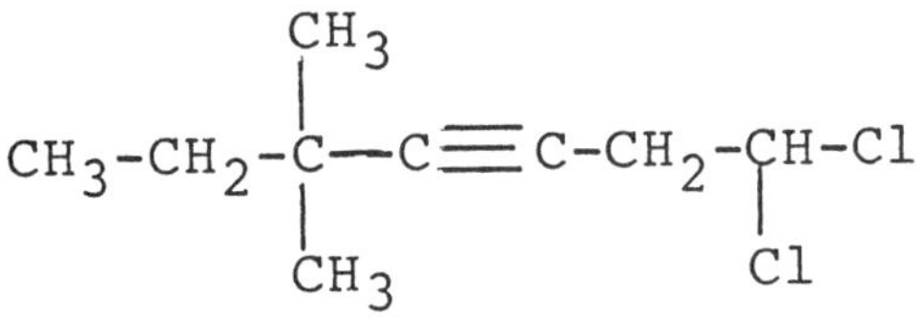

ANSWERS TO PROBLEM 18-9

(a) 5-bromo-2-pentyne (Give the lowest possible number to the triple bond.)

(b) 5,5-dimethyl-3-heptyne

(c) 1,1-dichloro-5,5-dimethyl-3-heptyne

(Give the lowest possible number to the triple bond by numbering from right to left in the chain.)

If your answers are correct, proceed to Problem 18-10. If your answers are incorrect, review Section 18-5, nomenclature Example 18-7 and Study Exercise 18-9) in your text.

Proceed to Problem 18-10.

PROBLEM 18-10

Write the structural formula for each of the following compounds:

(a) 1-hexyne

_____

(b)   3-octyne

(c)   1-bromo-4,4-dimethyl-2-heptyne

ANSWERS TO PROBLEM 18-10

(a) $H-C \equiv C-CH_2-CH_2-CH_2-CH_3$

(b) $CH_3-CH_2-C \equiv C-CH_2-CH_2-CH_2-CH_3$

(c)
$$Br-CH_2-C \equiv C-\overset{\overset{\displaystyle CH_3}{|}}{\underset{\underset{\displaystyle CH_3}{|}}{C}}-CH_2-CH_2-CH_3$$

If your answers are correct, proceed to Problem 18-11. If your answers are incorrect, review Section 18-5, nomenclature (Example 18-8 and Study Exercise 18-10) in your text.

Proceed to Problem 18-11.

PROBLEM 18-11

Complete and balance the reaction equation for each of the following chemical reactions:

(a) $CH_3-\overset{\overset{\displaystyle}{|}}{\underset{\underset{\displaystyle CH_3}{|}}{CH}}-C \equiv CH + Br_2 \quad ---->$
       excess in
       CCl_4

(b) $CH_3-CH_2-C \equiv C-\overset{\overset{\displaystyle}{|}}{\underset{\underset{\displaystyle CH_3}{|}}{CH}}-CH_3 + Cl_2 (g) \quad ---->$
       excess

319

ANSWERS TO PROBLEM 18-11

(a) 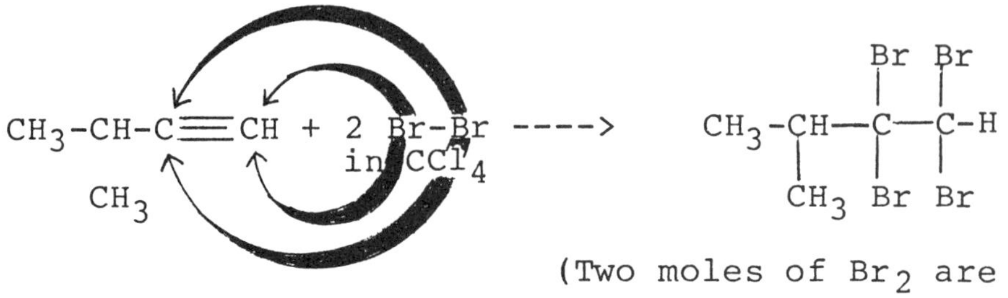 $CH_3-CH-C\equiv CH + 2\ Br-Br \xrightarrow{in\ CCl_4}$

$CH_3-CH-\underset{\underset{Br}{|}}{\overset{\overset{Br}{|}}{C}}-\underset{\underset{Br}{|}}{\overset{\overset{Br}{|}}{C}}-H$

with CH_3 substituent

(Two moles of Br_2 are used. Two of the three bonds in the triple bond break. None of the C-C single bonds break.)

(b) 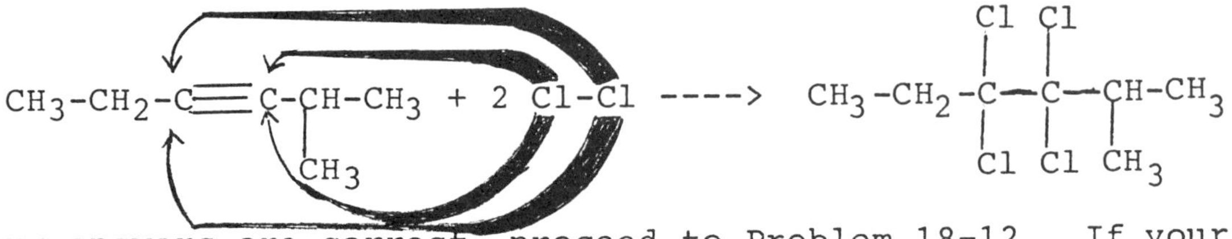 $CH_3-CH_2-C\equiv C-CH-CH_3 + 2\ Cl-Cl \longrightarrow CH_3-CH_2-\underset{\underset{Cl}{|}}{\overset{\overset{Cl}{|}}{C}}-\underset{\underset{Cl}{|}}{\overset{\overset{Cl}{|}}{C}}-\underset{\underset{CH_3}{|}}{CH}-CH_3$

with CH_3 substituent

If your answers are correct, proceed to Problem 18-12. If your answers are incorrect, review Section 18-5, uses and reactions of alkynes (Study Exercise 18-11) in your text.

Proceed to Problem 18-12.

PROBLEM 18-12

Write the IUPAC name for each of the following compounds:

(a) 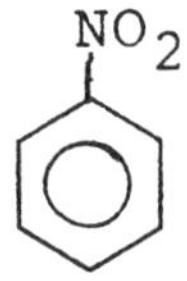 NO_2

_____

(b) 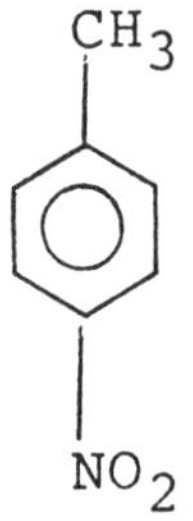 CH_3 / NO_2

_____

(c) 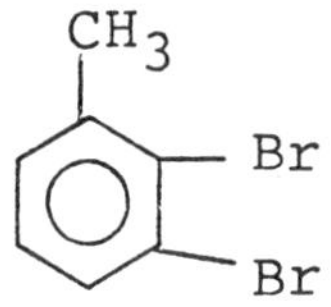 CH_3 / Br / Br

_____

320

ANSWERS TO PROBLEM 18-12

(a) nitrobenzene

(b) p-nitrotoluene

(c) 2,3-dibromotoluene

If your answers are correct, proceed to Problem 18-13. If your answers are incorrect, review Section 18-6, nomenclature (Example 18-9 and Study Exercise 18-12) in your text.

Proceed to Problem 18-13.

PROBLEM 18-13

Write the structural formula for each of the following compounds:

(a) p-bromoethylbenzene

_____

(b) 3,4-dinitrotoluene

_____

(c) 2-chloro-3,4-dinitrotoluene

_____

ANSWERS TO PROBLEM 18-13

(a)

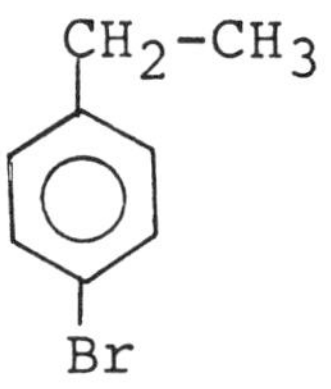

(b)

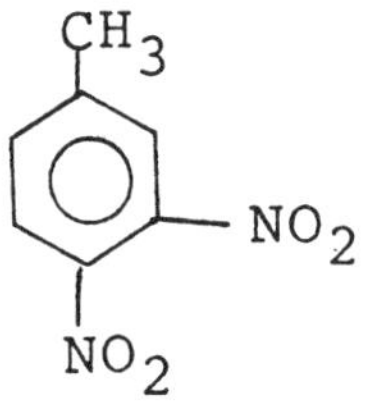

(c)

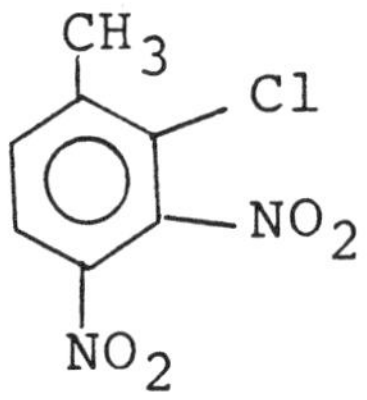

If your answers are correct, proceed to Problem 18-14. If your
answers are incorrect, review Section 18-6 nomenclature
Example 18-10 and Study Exercise 18-13) in your text.

Proceed to Problem 18-14.

PROBLEM 18-14

Complete and balance the reaction equation for each of the following chemical reactions:

(a)

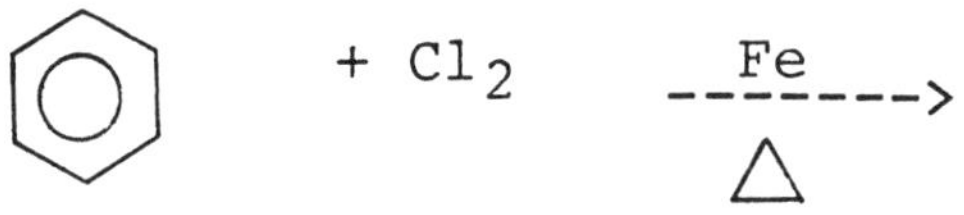

$+ Cl_2$

$\xrightarrow[\triangle]{Fe}$

(b)

$+ Br_2$

$\xrightarrow[\triangle]{Fe}$

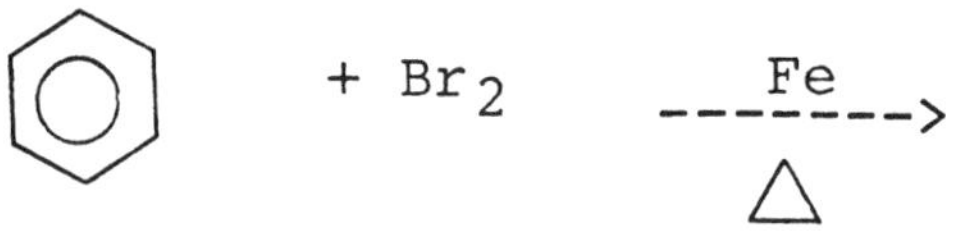

ANSWERS TO PROBLEM 18-14

(a)

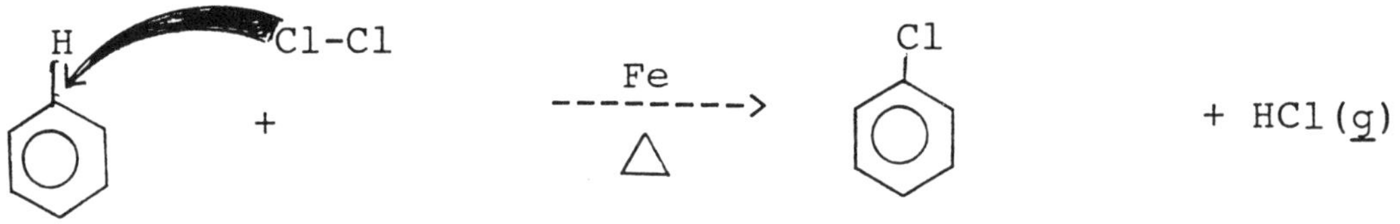

$+$

$\xrightarrow[\triangle]{Fe}$

$+ HCl(\underline{g})$

(b)

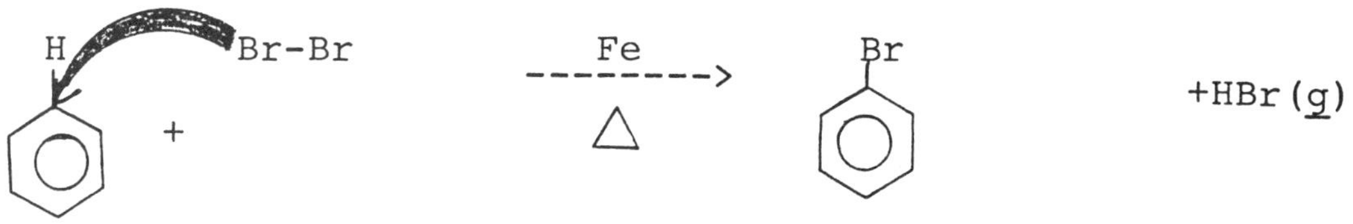

+HBr(g)

If your answers are correct, proceed to Problem 18-15. If your answers are incorrect, review Section 18-6, uses and reactions of aromatic compounds (Study Exercise 18-14) in your text.

Proceed to Problem 18-15.

PROBLEM 18-15

Identify the hydrocarbon derivatives as (1) organic halides, (2) alcohols, (3) ethers, (4) aldehydes, (5) ketones, (6) carboxylic acids, (7) esters, (8) amines, or (9) amides, in each of the following compounds by circling each group and labeling it.

(a)

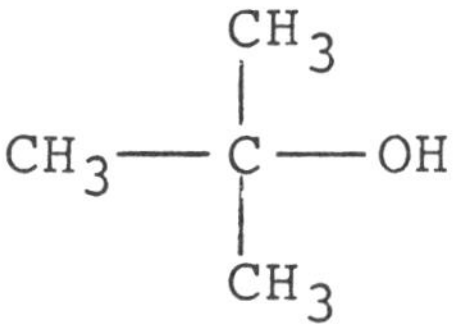

$$CH_3-\overset{\overset{\displaystyle CH_3}{|}}{\underset{\underset{\displaystyle CH_3}{|}}{C}}-OH$$

(b) 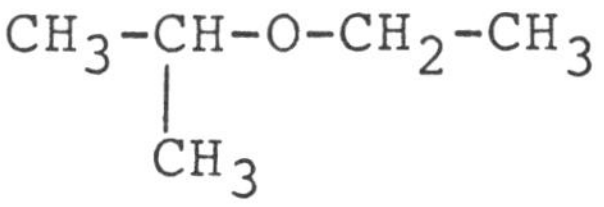 $CH_3-\overset{\overset{\displaystyle |}{CH}}{\underset{\underset{\displaystyle CH_3}{|}}{}}-O-CH_2-CH_3$

(c)

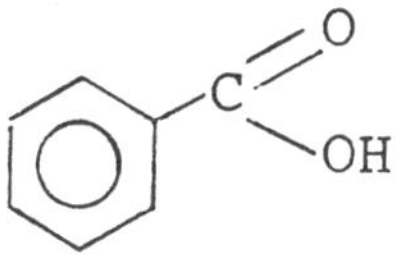

(d)

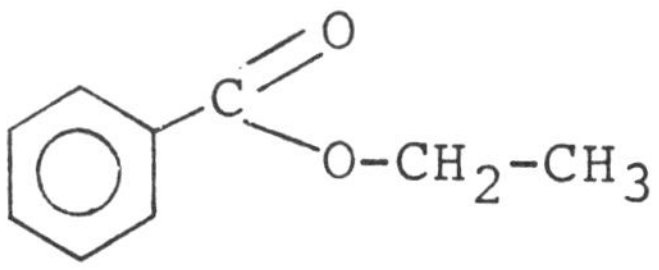

(e)

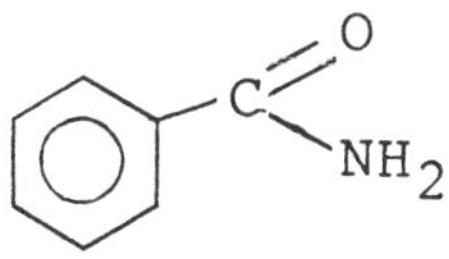

ANSWERS AND SOLUTIONS TO PROBLEM 18-15

(a)   alcohol,

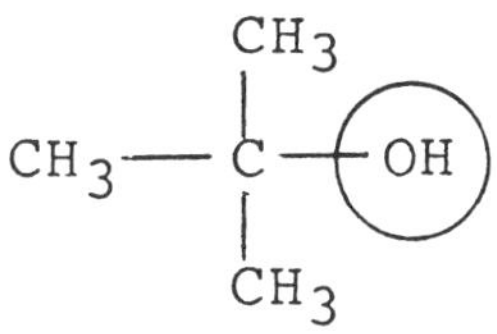

(b)   ether, $CH_3-CH-O-CH_2-CH_3$
                      $|$
                      CH_3

(c)   carboxylic acid,

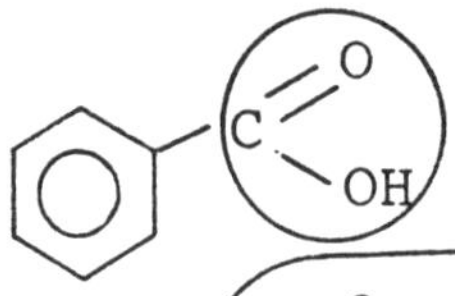

ϕ—$C$$\overset{\displaystyle O}{\diagdown}$$OH$

(d)   ester,

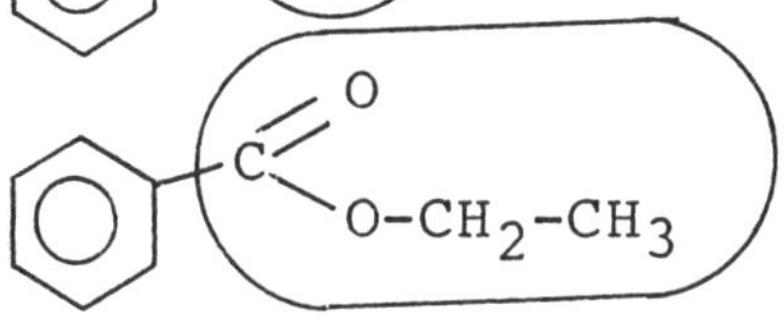

ϕ—$C$$\overset{\displaystyle O}{\diagdown}$$O-CH_2-CH_3$

(e)   amide,

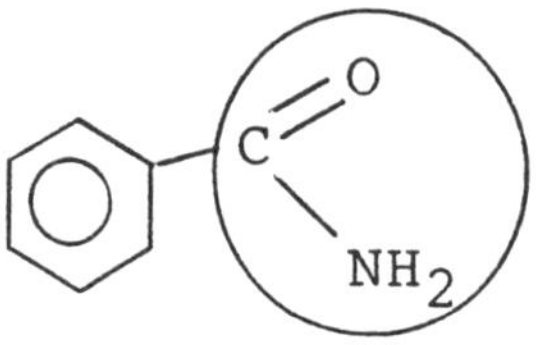

ϕ—$C$$\overset{\displaystyle O}{\diagdown}$$NH_2$

If your answers are correct, you have completed this chapter. If your answers are incorrect, review Section 18-7 (Study Exercise 18-15) in your text.

Now take the sample quizzes to see if you have mastered the material in this chapter.

SAMPLE QUIZZES

The quizzes divide this chapter into three parts: Quiz #18A (Sections 18-1 to 18-3), Quiz #18B (Sections 18-4 to 18-6),and Quiz #18C (Section 18-7).

Quiz #18A (Sections 18-1 to 18-3)

1. Write condensed structural formulas for the isomers of C_6H_{14}. There are five isomers.

2. Write the IUPAC name for each of the following compounds:

(a) $CH_2-CH-CH_2-CH_2-CH_2-CH_3$
         $|$ $|$
         Cl CH_3

(b) 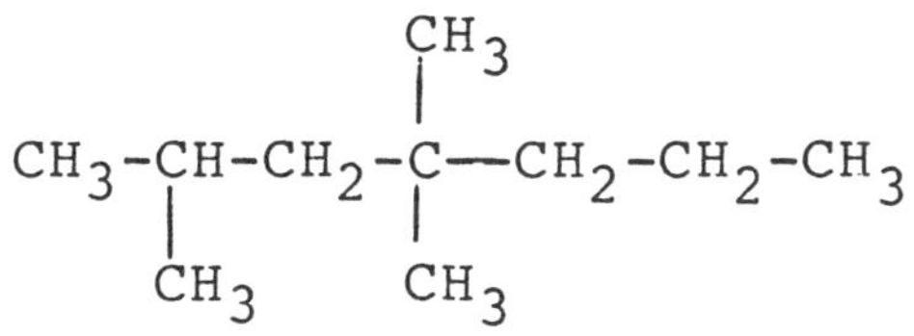 $CH_3-CH-CH_2-\overset{\overset{\displaystyle CH_3}{|}}{\underset{\underset{\displaystyle CH_3}{|}}{C}}-CH_2-CH_2-CH_3$
$\underset{\displaystyle CH_3}{|}$

_____

(c) $CH_3-\overset{\overset{\displaystyle CH_3}{|}}{\underset{\underset{\displaystyle CH_3}{|}}{C}}-CH_2-CH_2-\underset{\underset{\displaystyle CH_2-CH_2-CH_3}{|}}{CH}-CH_2-CH_3$

_____

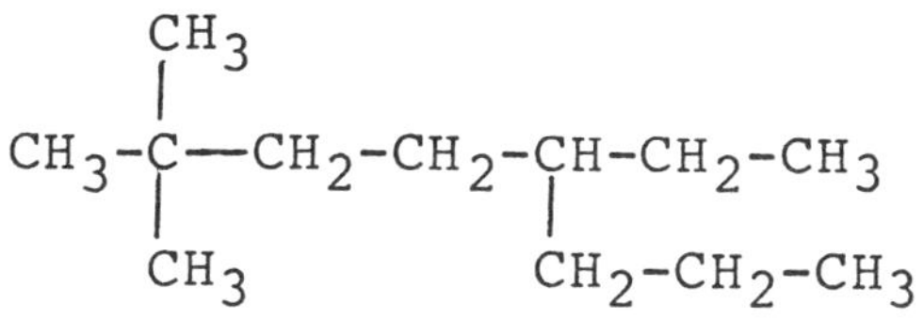

(d)

_____

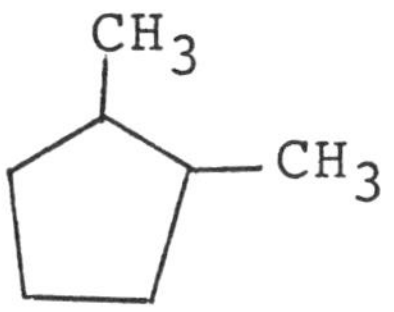

3. Write the structural formula for each of the following compounds:

(a)   isopropyl bromide

_____

(b)   2,2-dimethylbutane

_____

(c)   2-bromobutane

_____

(d)   methylcyclobutane

_____

325

Answers for Quiz #18A

1. CH$_3$-CH$_2$-CH$_2$-CH$_2$-CH$_2$-CH$_3$

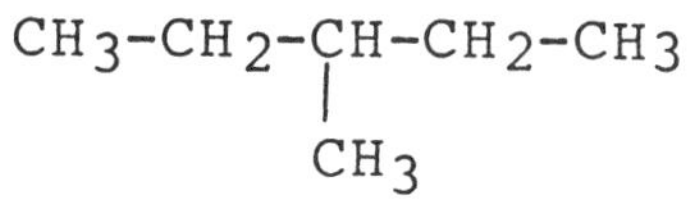

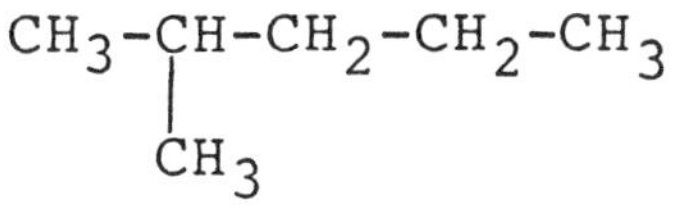

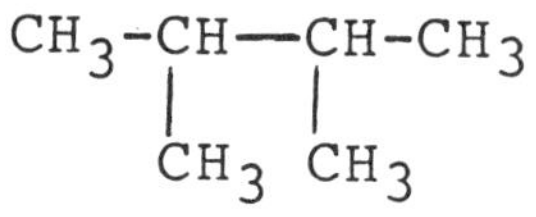

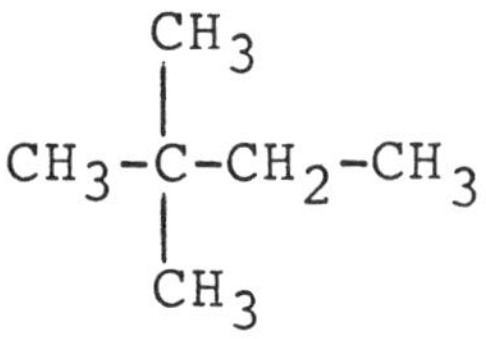

CH$_3$-CH$_2$-CH-CH$_2$-CH$_3$
                |
                CH$_3$

CH$_3$-CH-CH$_2$-CH$_2$-CH$_3$
        |
        CH$_3$

CH$_3$-CH—CH-CH$_3$
        | |
        CH$_3$ CH$_3$

         CH$_3$
          |
CH$_3$-C-CH$_2$-CH$_3$
          |
         CH$_3$

2. (a) 1-chloro-2-methylhexane; (b) 2,4,4-trimethylheptane;

(c) 5-ethyl-2,2-dimethyloctane;

(d) 1,2-dimethylcyclopentane

3. (a) 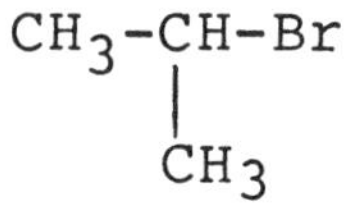 CH$_3$-CH-Br   (b) 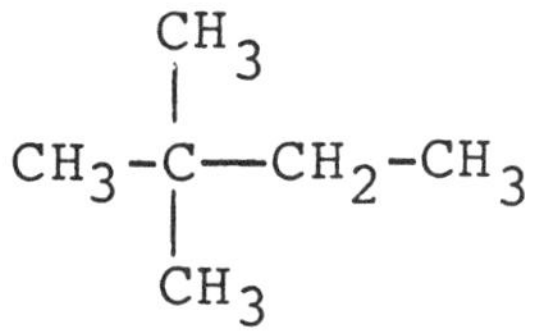   CH$_3$
              | |
              CH$_3$ CH$_3$-C—CH$_2$-CH$_3$
                                           |
                                          CH$_3$

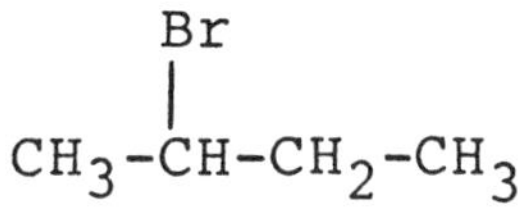

(c)   Br   (d) 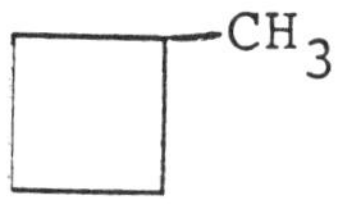
             |
    CH$_3$-CH-CH$_2$-CH$_3$ ┌───┐-CH$_3$
                                      │   │
                                      └───┘

Quiz #18B (Sections 18-4 to 18-6)

1. Write the IUPAC name for each of the following compounds:

(a) CH$_3$-CH$_2$-CH=CH$_2$

_____

(b) 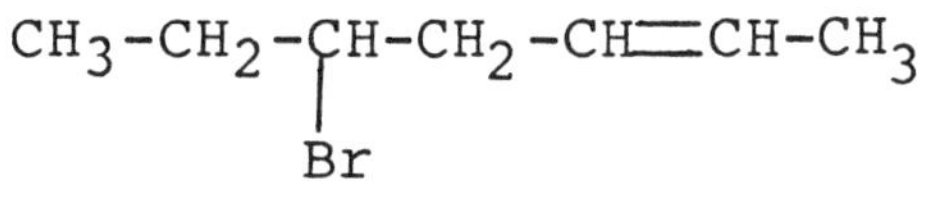 CH$_3$-CH$_2$-CH-CH$_2$-CH=CH-CH$_3$
                 |
                 Br

_____

(c) 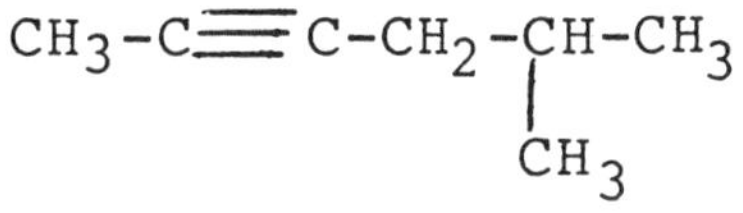 CH$_3$-C≡C-CH$_2$-CH-CH$_3$
                       |
                       CH$_3$

_____

326

(d)

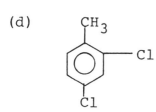

2. Write the structural formula for each of the following compounds:

(a) 4-methyl-1-pentene

(b) 2-heptyne

(c) cyclohexene

(d) m-nitrochlorobenzene

3. Complete and balance the reaction equation for the following chemical reaction.

$$CH_3-\underset{\underset{CH_3}{|}}{CH}-CH\equiv CH_2 \quad + \quad Br_2 \quad ----> \quad in\ CCl_4$$

Answers for Quiz #18B

1. (a) 1-butene; (b) 5-bromo-2-heptene; (c) 5-methyl-2-hexyne:

(d) 2,4-dichlorotoluene

2. (a) $CH_2\!=\!CH\text{-}CH_2\text{-}\underset{\underset{\displaystyle CH_3}{|}}{CH}\text{-}CH_3$

   (b) $CH_3\text{-}C\!\equiv\!C\text{-}CH_2\text{-}CH_2\text{-}CH_2\text{-}CH_3$

   (c)

   (d)

3. $CH_3\text{-}\underset{\underset{\displaystyle CH_3}{|}}{CH}\text{-}CH\!=\!CH_2$ $+$ Br_2 $\text{-----}\!>$ $\qquad$ $CH_3\text{-}\underset{\underset{\displaystyle CH_3}{|}}{CH}\text{-}\underset{\underset{\displaystyle Br}{|}}{CH}\text{-}\underset{\underset{\displaystyle Br}{|}}{CH_2}$

   $\qquad\qquad\qquad$ in CCl_4

Quiz #18C (Section 18-7)

1. Circle the functional group in each of the following molecules and label it.

   (a)
   $CH_3\text{-}CH_2\text{-}C\!\!\begin{array}{c}{\diagup O}\\[-2pt]{\diagdown CH_3}\end{array}$

   (b)
   $CH_3\text{-}CH_2\text{-}C\!\!\begin{array}{c}{\diagup O}\\[-2pt]{\diagdown O\text{-}CH_3}\end{array}$

2. Identify the hydrocarbon derivatives as (1) organic halides, (2) alcohols, (3) ethers, (4) aldehydes, (5) ketones, (6) carboxylic acids, (7) esters, (8) amines, or (9) amides in each of the following compounds by circling each group and labeling it:

   (a)

   (b)

3. Draw the functional group associated with each of the following classes of hydrocarbon derivatives.

   (a) carboxylic acid   (b) aldehyde

328

1. (a) ketone, 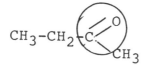 ; (b) ester,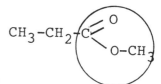

2. (a) aldehyde

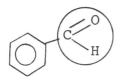

(b) ketone

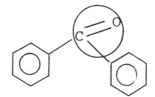

3. (a) ;

329

CHAPTER 19

NUCLEAR CHEMISTRY

In ordinary chemical equations, valence electrons of atoms are gained, lost, or shared with no changes in the nucleus of the atom. We will now consider changes in the nucleus that result in one element being changed to an entirely different element. Reactions that change the nucleus of the atom are nuclear reactions; for these reactions we write nuclear equations. Radioactivity accompanies nuclear reactions. Radioactivity is the spontaneous emissions (radiations) accompanying changes in the nuclei of atoms.

SELECTED TOPICS

1. The five nuclear emissions (radiations) associated with nuclear reactions are:

        (1)   alpha (α) particles
        (2)   beta (β) particles
        (3)   gamma (γ) rays
        (4)   positron particles or positrons
        (5)   neutrons

Alpha particles are nuclei of helium-4 atoms. They have a 2^+ charge, possess an atomic number of 2 and a mass number of 4 and are written $^4_2$He. The 2^+ charge is normally omitted because the helium nucleus $(^4_2$He$^{2+})$ will rapidly pick up two electrons from some other atom nearby or from electrons in the atmosphere to form the neutral helium atom $(^4_2$He$)$. The velocity of an alpha particle is about 0.1 times the velocity of light $(3.00 \times 10^{10}$ cm/s$)$. Alpha particles have very low penetration power and can be stopped by a thin sheet of paper. They have a large ionizing effect on gases because of the 2^+ charge on the nucleus.

Beta particles are identical to electrons. They have a unit negative charge and come from the nucleus when a neutron is changed to a proton (n ---> p + β). A beta particle is written $^0_{-1}$e with the -1 representing the atomic number and 0 the mass number, because electrons have negligible mass. They have a velocity of about 0.9 times the velocity of light. They have a greater penetrating power

330

than alpha particles, but can be stopped by a sheet of aluminum 1 cm thick. They have a small ionizing effect on gases.

Gamma rays are rays of light energy of very short wavelength, similar to x-rays. They have no charge or mass and are emitted from the nucleus as a result of an internal change in the nucleus that produces energy. They often accompany a particle loss (alpha particle, beta particle, positron, or neutron). The velocity of gamma rays is equal to the velocity of light. They have very high energy and high penetrating power, and readily penetrate the body. Lead or concrete is needed to stop these rays. Gamma rays have an even smaller ionizing effect on gases than beta particles. Alpha and beta particles and gamma rays are the radiations associated with natural radioactivity. See Section 19-2 in your text.

Positron particles or positrons are identical to positive electrons and have a unit positive charge. They come from the nucleus when a proton is transformed into a neutron

(p ---> positron + n). A positron is written $^{0}_{+1}e$, with the 1 representing the atomic number and 0 the mass number, because electrons have negligible mass. Positrons are similar to beta particles in velocity and ionizing effect, but they have very low penetrating power.

Neutrons ($^{1}_{0}n$) have variable velocities (1×10^{-5} to 1×10^{-1} times the velocity of light). They have relatively high penetrating power, but no ionizing effect, as they have no charge. Positron particles or positrons and neutrons are the radiations associated with artificial radioactivity. See Section 19-3 in your text.

Table 19-1 in your text summarizes these radiations.

2. We will now use these nuclear radiations to write nuclear equations for nuclear reactions. Table 19-1 in your text gives the names of nuclear radiations used in nuclear equations and their symbols. You must know these symbols. In addition to the five nuclear radiations previously mentioned and included in Table 19-1, this table also lists other particles used in nuclear reactions, the proton

($^{1}_{1}H$), the electron ($^{0}_{-1}e$), the deuteron ($^{2}_{1}H$, an isotope of hydrogen), and the triton ($^{3}_{1}H$, another isotope of hydrogen). In nuclear equations the sum of the atomic numbers and the sum of the mass numbers in the reactants must equal their corresponding sum of atomic numbers and the sum of the mass numbers in the products. In nuclear reactions, there is some loss in mass; this loss is converted to energy according to Einstein's equation, $E = mc^2$. This loss in mass is not shown in nuclear equations in the mass number because it is a small fraction of an atomic mass unit. Gamma rays are also not shown in nuclear equations because their loss does not alter the nuclear equation. Gamma rays often accompany a particle loss. Sections 19-2 and 19-3 in your text gives examples of nuclear reactions.

3. The <u>half-life</u> of a radioactive isotope is the amount of time required for one-half of a given mass of a radioactive element to decay. This value is constant for a given radioactive isotope. It is independent of external conditions, such as temperature, pressure, or the oxidation state of the element. The half-life of sulfur-35 is 88 days. This means if we now have 1.000 mg, in 88 days we will have 0.500 mg of sulfur-35 (one-half of 1.000 mg); in 176 days (88 days + 88 days = 176 days) we will have 0.250 mg of sulfur-35 (one-half of 0.500 mg); and in 264 days (176 days + 88 days = 264 days) we will have 0.125 mg of sulfur-35 (one-half of 0.250 mg). Section 19-4 in your text gives various examples of problems using half-lives. See Section 19-4 in your text.

PROBLEM 19-1

Complete and balance a nuclear chemical equation for each of the following nuclear reactions. (Refer to the periodic table on the inside front cover of your text to answer these questions.)

(a) Actinium (Ac)-228 decays by beta emission

(b) Bismuth-210 decays by beta emission

(c) Radon (Rn)-219 decays by alpha emission

(d) Radium (Ra)-223 decays by alpha emission

ANSWERS AND SOLUTIONS TO PROBLEM 19-1

(a) From the periodic table, the atomic number for actinium (Ac) is 89. A beta particle is $_{-1}^{0}e$. Decay means "to give off"; actinium is on the reactants side of the equation, and the beta particle plus some other element is on the products side. We can write an incomplete equation as follows:

$$^{228}_{89}\text{Ac} \; \text{---}\!\!> \; ^{0}_{-1}\text{e} \; + \; ?$$

For the atomic number in the reactant to equal the sum of the atomic numbers in the products, the atomic number of the new element must be 90.

$$89 = -1 + x$$

$$x = 90$$

The mass number of the new element is 228. Refer again to the periodic table and look for the symbol of the element with an <u>atomic number</u> of 90. The symbol is Th; from the list of elements and their symbols (inside front cover), the name of the element is thorium. Therefore, the equation for this nuclear reaction is

$$^{228}_{89}\text{Ac} \; \text{---}\!\!> \; ^{0}_{-1}\text{e} \; + \; ^{228}_{90}\text{Th}$$

(b) From the periodic table, the atomic number for bismuth is 83. The symbol for a beta particle is $^{0}_{-1}\text{e}$. We can write an incomplete equation as follows:

$$^{210}_{83}\text{Bi} \; \text{---}\!\!> \; ^{0}_{-1}\text{e} \; + \; ?$$

The atomic number for the new element is 84.

$$83 = -1 + x$$

$$x = 84$$

The mass number for the new element is 210. Refer again to the periodic table for the symbol of the element with an atomic number of 84. The symbol is Po and the name of the element is polonium. Therefore, the equation for this nuclear reaction is

$$^{210}_{83}\text{Bi} \; \text{---}\!\!> \; ^{0}_{-1}\text{e} + ^{210}_{84}\text{Po}$$

(c) From the periodic table, the atomic number for radon (Rn) is 86. The symbol for an alpha particle is $^{4}_{2}\text{He}$. We can write an incomplete equation as follows:

$$^{219}_{86}\text{Rn} \; \text{---}\!\!> \; ^{4}_{2}\text{He} + ?$$

The atomic number for the new element is 84.

$$86 = 2 + x$$

$$x = 84$$

We can apply the same method to the mass number. The mass number of the new element is 215.

$$219 = x + 4$$

$$x = 215$$

Refer again to the periodic table for the symbol of the element with an atomic number of 84. The symbol is Po and the name of the element is polonium. Therefore, the equation for the nuclear reaction is

$$^{219}_{86}Rn \longrightarrow {}^{4}_{2}He + {}^{215}_{84}Po$$

(d) From the periodic table, the atomic number for radium (Ra) is 88. The symbol for an alpha particle is $^{4}_{2}He$. We can write an incomplete equation as follows:

$$^{223}_{88}Ra \longrightarrow {}^{4}_{2}He + ?$$

The atomic number for the new element is 86.

$$88 = 2 + x$$

$$x = 86$$

The mass number for the new element is 219.

$$223 = 4 + x$$

$$x = 219$$

Refer again to the periodic table for the symbol of the element with an atomic number of 86. The symbol is Rn and the name of the element is radon. Therefore, the equation for the nuclear reaction is

$$^{223}_{88}Ra \longrightarrow {}^{4}_{2}He + {}^{219}_{86}Rn$$

If your answers are correct, proceed to Problem 19-2. If your answers are incorrect, review Section 19-2 (Examples 19-1 and 19-2 and Study Exercise 19-1) in your text.

Proceed to Problem 19-2.

PROBLEM 19-2

Bombardment of strontium-86 with a deuteron results in the formation of another element and the release of an alpha particle for each strontium atom. Complete and balance a nuclear chemical equation for this nuclear reaction.

ANSWER AND SOLUTION TO PROBLEM 19-2

Use the periodic table to determine the atomic number of strontium (38). The symbol for deuteron is ^2_1H and the symbol for an alpha particle is ^4_2He. This information allows us to write an incomplete equation:

$$^{86}_{38}Sr \ + \ ^2_1H \ ---> \ ^4_2He \ + \ ?$$

The atomic number of the new element is 37.

$$38 + 1 = 2 + x$$

$$x = 37$$

The mass number of the new element is 84.

$$86 + 2 = 4 + x$$

$$x = 84$$

Refer again to the periodic table for the symbol of the element with an atomic number of 37. The symbol is Rb and the name is rubidium. Therefore, the equation for this nuclear reaction is

$$^{86}_{38}Sr \ + \ ^2_1H \ ---> \ ^4_2He \ + \ ^{84}_{37}Rb$$

If your answer is correct, proceed to Problem 19-3. If your answer is incorrect, review Section 19-3 (Example 19-3 and 19-4 and Study Exercise 19-2) in your text.

Proceed to Problem 19-3.

PROBLEM 19-3

An isotope of silver, silver-111, has a half-life of 7.6 days. If 1.000 mg of silver-111 disintegrates over a period of 22.8 days, how many milligrams of silver-111 will remain?

ANSWER AND SOLUTION TO PROBLEM 19-3

0.125 mg At this time (0 days), we have 1.000 mg of the isotope; after 7.6 days we shall have 1/2 of the original or 0.500 mg; and after 15.2 days we shall have 1/2 of 0.500 mg, or 0.250 mg. Finally, after 22.8 days, we shall have 1/2 of 0.250 mg, or 0.125 mg.

Add half-life each time	0 days	1.000 mg	Divide the amount by two each time
	7.6 days	0.500 mg	
	15.2 days	0.250 mg	
	22.8 days	0.125 mg	

If your answer is correct, proceed to Problem 19-4. If your answer is incorrect, review Section 19-4 (Example 19-5) in your text.

Proceed to Problem 19-4.

PROBLEM 19-4

The half-life of radium-226 is 1622 years. How many years will it take for a sample of radium-226 to decrease from a mass of 280.0 mg to a mass of 70.0 mg?

ANSWERS AND SOLUTION TO PROBLEM 19-4

3244 years

In 0 years, we have 280.0 mg, in 1622 years we shall
have 1/2 of 280.0 mg or 140.0 mg, and in 3244 years
(1622 years + 1622 years = 3244 years) we shall have
70.0 mg.

Divide the amount by two each time	280.0 mg	0 years	Add half-life each time
	140.0 mg	1622 years	
	70.0 mg	3244 years	

If your answer is correct, you have completed this chapter. If
your answer is incorrect, review Section 19-4 (Example 19-6 and
Study Exercise 19-3) in your text.

These problems conclude the chapter on nuclear chemistry.

Now take the following quiz to see if you have mastered the
material in this chapter.

SAMPLE QUIZ

Quiz #19   You may use the periodic table.

1. Complete and balance a nuclear chemical equation for the
    following nuclear reactions:

    (a) Krypton(Kr)-87 decays by beta emission

    (b) Gadolinium(Gd)-152 decays by alpha emission

    (c) Palladium(Pd)-108 is bombarded with a high speed alpha
         particle; a proton is emitted in the process

2. An isotope of xenon, xenon-135, has a half-life of 9.2
    hours. If 3.000 mg of xenon-135 disintegrates over a period
    of 27.6 hours, how many milligrams of xenon-135 will
    remain?

3. The half-life of iodine-120 is 1.3 hours. How many hours will it take a sample of iodine-120 to decrease from a mass of 20.00 mg to a mass of 2.50 mg?

Answers and Solutions to Quiz #19

1. (a) $^{87}_{36}Kr \longrightarrow \ ^{0}_{-1}e \ + \ ^{87}_{37}Rb$

    (b) $^{152}_{64}Gd \longrightarrow \ ^{4}_{2}He \ + \ ^{148}_{62}Sm$

    (c) $^{108}_{46}Pd \ + \ ^{4}_{2}He \longrightarrow \ ^{1}_{1}H \ + \ ^{111}_{47}Ag$

2.
0 hours	3.000 mg	
9.2 hours	1.500 mg	
18.4 hours	0.750 mg	
27.6 hours	0.375 mg	Answer

3.
20.00 mg	0 hours	
10.00 mg	1.3 hours	
5.00 mg	2.6 hours	
2.50 mg	3.9 hours	Answer

338

Review Exam #6 (Chapters 16 to 19) [Answers are given in ( ) to the right of the questions.]

You may use the periodic table.

1-3 Given the following unbalanced equation:

$$FeCl_3 + H_2S \longrightarrow FeCl_2 + S + HCl$$

1. The increase or decrease in oxidation number (ox. no.) of Fe in $FeCl_3$ is

   A. increase 1 ox. no.   (B)

   B. decrease 1 ox. no.

   C. increase 2 ox. no.

   D. decrease 2 ox. no.

   E. decrease 3 ox. no.

2. The coefficients in the balanced equation are:

   A. $1 + 1 \longrightarrow 1 + 1 + 1$ (D)

   B. $2 + 1 \longrightarrow 1 + 1 + 2$

   C. $1 + 2 \longrightarrow 1 + 2 + 2$

   D. $2 + 1 \longrightarrow 2 + 1 + 2$

   E. $2 + 1 \longrightarrow 2 + 6$

3. The oxidizing agent is:

   A. $FeCl_3$ (A)

   B. H_2S

   C. $FeCl_2$

   D. S

   E. HCl

339

4-6 Given the following unbalanced equation:

$$Cu + NO_3^- \longrightarrow Cu^{2+} + NO \text{ in acid solution}$$

4. The balanced oxidation half-reaction is:

    A. $NO_3^- \longrightarrow NO + 1 e^-$ (D)

    B. $NO_3^- + 2 H^+ + 1 e^- \longrightarrow NO + H_2O$

    C. $NO_3^- + 4 H^+ + 3 e^- \longrightarrow NO + 2 H_2O$

    D. $Cu \longrightarrow Cu^{2+} + 2 e^-$

    E. $Cu \longrightarrow Cu^{2+} + 1 e^-$

5. The coefficients in the balanced equation are:

    A. $3 + 2 + 8 H^+ \longrightarrow 3 + 2 + 4 H_2O$ (A)

    B. $1 + 1 \longrightarrow 1 + 2$

    C. $1 + 1 \longrightarrow 1 + 1 + 2 H_2O$

    D. $1 + 1 \longrightarrow 1 + 1$

    E. $3 + 2 \longrightarrow 3 + 2 + 4 H_2O$

6. The reducing agent is:

    A. Cu   (A)

    B. NO_3^-

    C. Cu^{2+}

    D. NO

340

7. Write the equilibrium expression for the following reaction:

$$4 \text{ H}_2(g) + \text{CS}_2(g) \rightleftarrows \text{CH}_4(g) + 2 \text{ H}_2\text{S}(g)$$

A. $K = \dfrac{[CH_4][H_2S]}{[H_2][CS_2]}$ (D)

B. $K = \dfrac{[CS_2][H_2]}{[CH_4][H_2S]}$

C. $K = \dfrac{[H_2]^4[CS_2]}{[CH_4][H_2S]^2}$

D. $K = \dfrac{[CH_4][H_2S]^2}{[H_2]^4[CS_2]}$

E. $K = \dfrac{[CH_4] + [H_2S]^2}{[H_2]^4 + [CS_2]}$

8. Given the following equilibrium reaction:

$$2 \text{ SO}_2(g) + \text{O}_2(g) \rightleftarrows 2 \text{ SO}_3(g) + \text{heat}$$

The equilibrium can be shifted to the right by

A. adding a catalyst   (C)

B. removing O_2 gas

C. increasing the pressure

D. increasing the temperature

E. adding SO_3 gas

9. Calculate the ionization constant (K_a) for a 0.200 $\underline{M}$ HA solution which is 5.50 % ionized.

A. 1.06×10^{-4} (mol/L)   (D)

B. 5.82×10^{-1} (mol/L)

C. 3.22×10^{-4} (mol/L)

D. 6.40×10^{-4} (mol/L)

E. 6.05×10^{-4} (mol/L)

10. Calculate the hydrogen ion concentration in moles per liter for a 0.100 $\underline{M}$ HA solution whose ionization constant ($K_{\underline{a}}$) is 4.00×10^{-7} (mol/L).

    A. 2.00×10^{-4} mol/L   (A)

    B. 6.32×10^{-4} mol/L

    C. 4.00×10^{-4} mol/L

    D. 6.32×10^{-3} mol/L

    E. 2.00×10^{-3} mol/L

11. Write the expression for the solubility product constant ($K_{\underline{sp}}$) of the slightly soluble electrolyte, Ag_2CrO_4.

    A. $K_{\underline{sp}} = [Ag^+][CrO_4^{2-}]$ (B)

    B. $K_{\underline{sp}} = [Ag^+]^2[CrO_4^{2-}]$

    C. $K_{\underline{sp}} = [Ag^+][CrO^{2-}]^4$

    D. $K_{\underline{sp}} = [Ag^+]^2[CrO^{2-}]^4$

    E. $K_{\underline{sp}} = [Ag^+][CrO_4^{2-}]^2$

12. Calculate the solubility product constant ($K_{\underline{sp}}$) if the solubility of silver chloride (AgCl) is 1.8×10^{-3} g/L at $25^{O}C$. (Atomic mass units: Ag = 107.9 amu, Cl = 35.5 amu)

    A. 4.2×10^{-2} (mol^2/L^2)   (E)

    B. 3.2×10^{-3} (mol^2/L^2)

    C. 3.2×10^{-6} (mol^2/L^2)

    D. 1.7×10^{-5} (mol^2/L^2)

    E. 1.7×10^{-10} (mol^2/L^2)

13. Calculate the solubility of zinc sulfide (ZnS) in grams per liter at 18°C if the solubility product constant (K_{sp}) of zinc sulfide is 1.2×10^{-28} (mol^2/L^2) at 18°C. (Atomic mass units: Zn = 65.4 amu, S = 32.1 amu)

A. 1.1×10^{-12} g/L (A)

B. 1.1×10^{-14} g/L

C. 1.4×10^{-28} g/L

D. 1.4×10^{-26} g/L

E. 1.09×10^{-14} g/L

14. There are three constitutional isomers of C_5H_{12}. Two of these isomers are:

(1) $CH_3-CH_2-CH_2-CH_2-CH_3$ and (2) 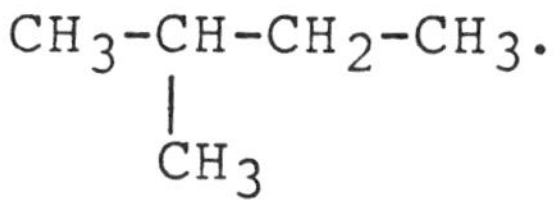 $CH_3-CH-CH_2-CH_3$. The third isomer is:

$$CH_3$$

A. 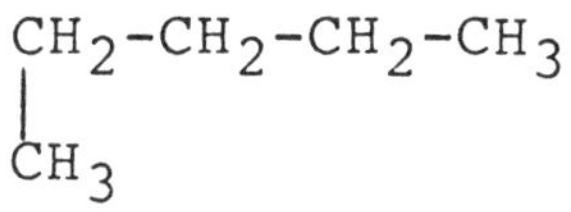 $CH_2-CH_2-CH_2-CH_3$ (E)
    $|$
    CH_3

B. 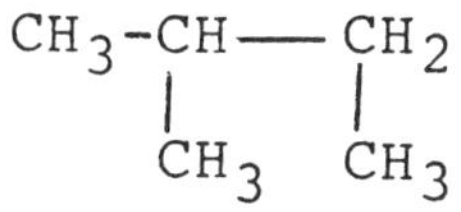 $CH_3-CH \longrightarrow CH_2$
        $|$ $|$
        CH_3 CH_3

C. 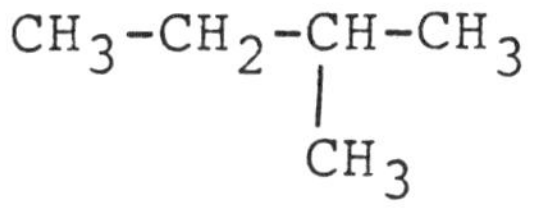 $CH_3-CH_2-CH-CH_3$
            $|$
            CH_3

D. 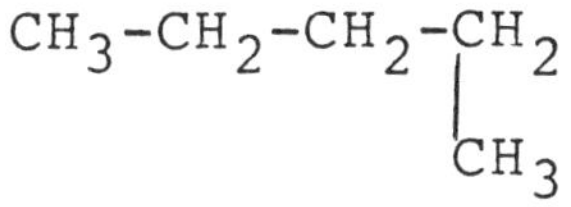 $CH_3-CH_2-CH_2-CH_2$
                $|$
                CH_3

E. 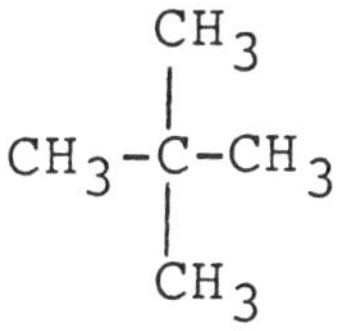 CH_3
        $|$
    CH_3-C-CH_3
        $|$
        CH_3

15. Write the IUPAC name for

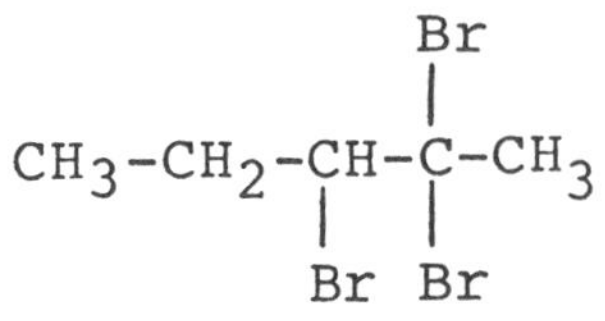

$$CH_3-CH_2-\underset{\underset{\displaystyle Br}{|}}{CH}-\underset{\underset{\displaystyle Br}{|}}{\overset{\overset{\displaystyle Br}{|}}{C}}-CH_3$$

A. 2,3-tribromopentane

B. 2,2,3-tribromopentane

C. 3,4,4-tribromopentane

D. 3,4-tribromopentane

E. 2,2,3-tribromobutane

(B)

16. Write the IUPAC name for:

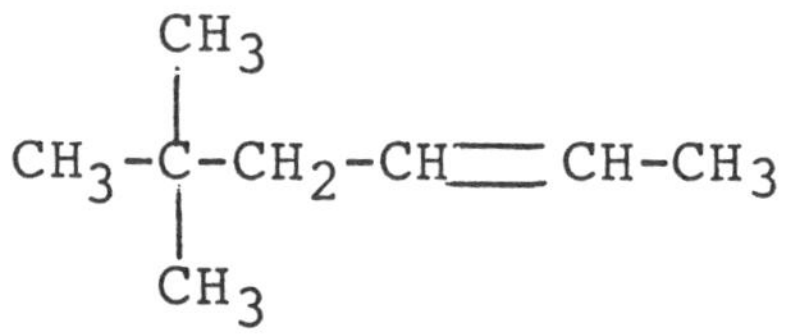

$$CH_3-\underset{\underset{\displaystyle CH_3}{|}}{\overset{\overset{\displaystyle CH_3}{|}}{C}}-CH_2-CH=CH-CH_3$$

A. 4-tert-butyl-2-hexene

B. 4-tert-butyl-2-butene

C. 2,2-dimethyl-4-hexene

D. 5-dimethyl-2-hexene

E. 5,5-dimethyl-2-hexene

(E)

17. Write the structural formula for 2,3-dichlorocyclopentene.

A. (B)

B.

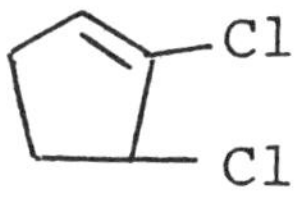

C. $CH_2=C(Cl)-CH(Cl)-CH_2-CH_3$

D.

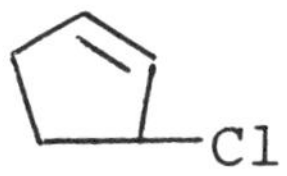

E.

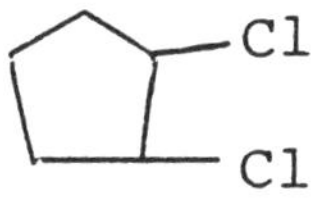

18. Complete and balance the reaction equation for the following chemical reaction:

$$CH_3-CH_2-\underset{\underset{CH_3}{|}}{C}=CH-CH_3 \; + \; Br_2 \; ----> \; \text{in } CCl_4$$

A.

   1 + 1 ---> $CH_3-CH_2-\underset{\underset{CH_3}{|}}{\overset{\overset{Br}{|}}{C}}-\overset{\overset{Br}{|}}{C}H-CH_3$ (A)

B.

   1 + 1 ---> $CH_3-CH_2-\overset{\overset{Br}{|}}{C}H-\overset{\overset{Br}{|}}{C}H-CH_3$

C.

   1 + 1 ---> $CH_3-CH_2-\underset{\underset{CH_3}{|}}{\overset{\overset{Br}{|}}{C}}-\overset{\overset{Br}{|}}{C}H_2$

D.

   1 + 1 ---> $CH_3-CH_2-\underset{\underset{CH_3}{|}}{\overset{\overset{H\;Br}{|\;\;|}}{C}}-\overset{\overset{Br}{|}}{C}H-CH_3$

E.

   1 + 2 ---> $CH_3-CH_2-\underset{\underset{Br}{|}}{\overset{\overset{Br}{|}}{C}}-\underset{\underset{Br}{|}}{\overset{\overset{Br}{|}}{C}}-CH_3$

19. Write the IUPAC name for:

$$CH_3-C\equiv C-\underset{\underset{CH_3}{|}}{\overset{\overset{CH_3}{|}}{C}}-CH_2-CH_3$$

A. 4-dimethyl-2-hexyne   (B)

B. 4,4-dimethyl-2-hexyne

C. 4,4-dimethyl-4-hexyne

D. 2-hexyne

E. 4,4-methyl-2-hexyne

346

20. Write the IUPAC name for:

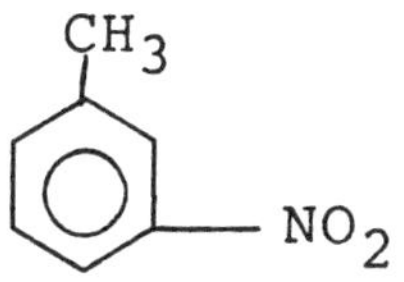

CH$_3$ on benzene ring with NO$_2$

A. $\underline{p}$-nitrotoluene   (D)

B. $\underline{p}$-nitrobenzene

C. $\underline{o}$-nitrotoluene

D. $\underline{m}$-nitrotoluene

E. $\underline{m}$-nitrobenzene

21. Write the structural formula for 2-bromo-4-$\underline{tert}$-butyltoluene.

A.   (C)

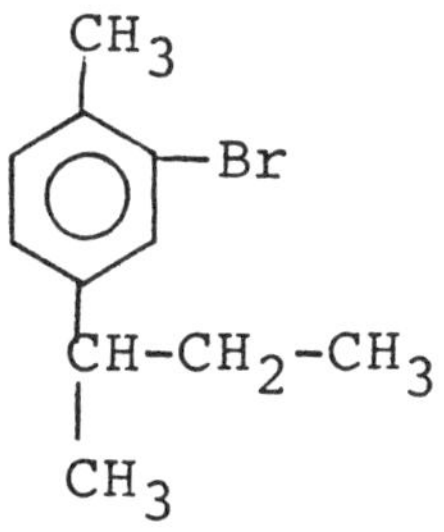

CH$_3$

Br

CH-CH$_2$-CH$_3$

CH$_3$

B.

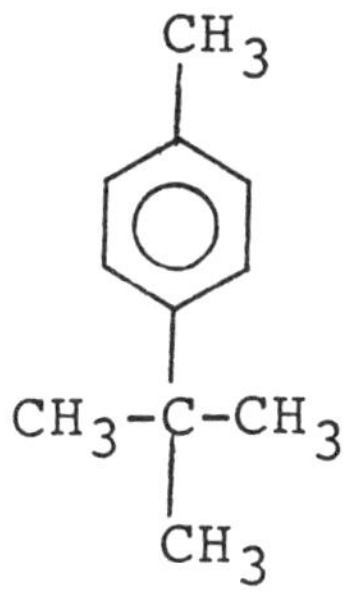

CH$_3$

CH$_3$-C-CH$_3$

CH$_3$

C.

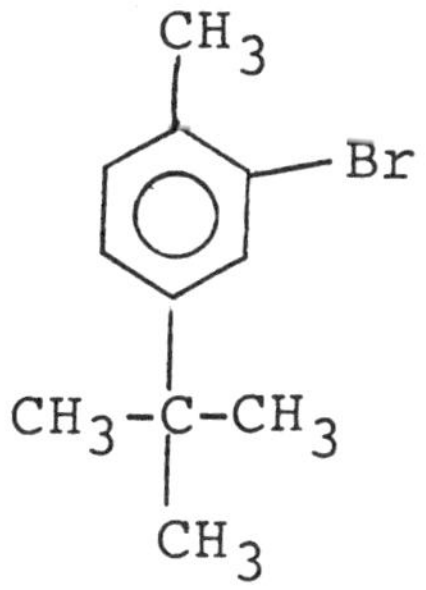

CH$_3$

Br

CH$_3$-C-CH$_3$

CH$_3$

D.

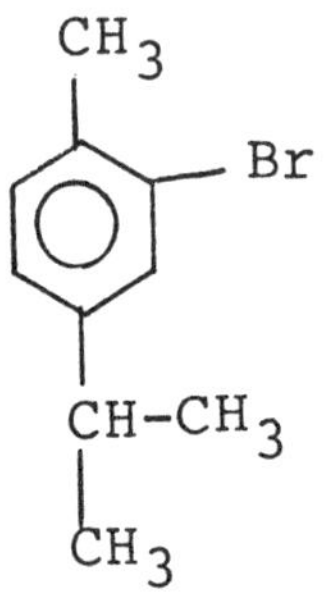

CH$_3$

Br

CH-CH$_3$

CH$_3$

E.

Br

CH$_3$-C-CH$_3$

CH$_3$

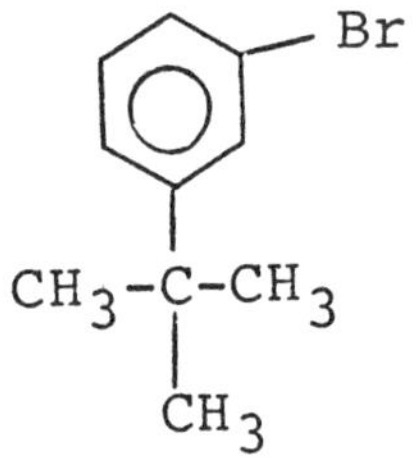

22. The following molecule is classified as:

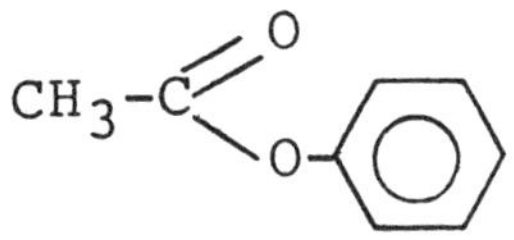

CH$_3$-C
O
O

A. a carboxylic acid   (C)

B. an aldehyde

C. an ester

D. a ketone

E. an alcohol

23. The following is the functional group for an amide:

(D)

A. 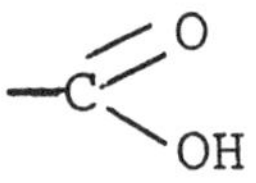
$-C$
O
OH

B. 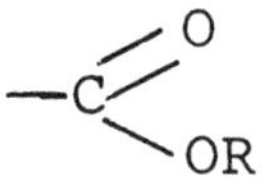
$-C$
O
OR

C. $-NH_2$

D. 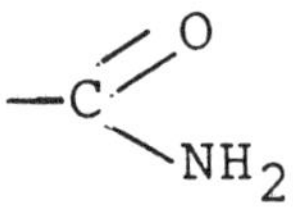
$-C$
O
NH_2

E. 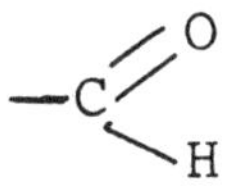
$-C$
O
H

348

24. Complete and balanced a nuclear chemical equation for the following nuclear reaction.

Iodine-130 decays by beta emission

A. $^{130}_{53}I + ^{0}_{+1}e \longrightarrow ^{130}_{54}Xe$ (C)

B. $^{130}_{53}I + ^{0}_{-1}e \longrightarrow ^{130}_{52}Te$

C. $^{130}_{53}I \longrightarrow ^{0}_{-1}e + ^{130}_{54}Xe$

D. $^{130}_{53}I \longrightarrow ^{0}_{+1}e + ^{130}_{52}Te$

E. $^{130}_{53}I \longrightarrow ^{1}_{1}H + ^{129}_{52}Te$

25. The half-life of barium-141 is 18 minutes. How many minutes will it take a sample of barium-141 to decrease from 1.000 mg to 0.125 mg?

A. 18 min   (B)

B. 54 min

C. 36 min

D. 144 min

E. 72 min

This review exam concludes your study of basic chemistry. We hope that your experience has been a pleasant one.